PRAISE FOR

MURRAY LONGE

"Murray distils decades of experience and complex concepts into an accessible roadmap."

KERI A. CLARKE

"Murray has a deep understanding of what he is discussing."

KROL- Amazon Reviewer

PRAISE FOR

HOW TO START YOUR START-UP

by MURRAY LONGE

"Like sitting down for an in-depth chat with a knowledgeable, experienced friend in the business. Practical advice based on real world practical success. The writing style makes it easy to read and absorb what might otherwise be quite complex ideas."

WORDSMITH - Amazon Reviewer

"A concise and practical guide for entrepreneurs, packed with clear insights, case studies, and actionable steps. An invaluable resource."

KERI A. CLARKE

"This book is a very well-written, thorough look at every aspect of starting a business. I highly recommend it, especially as it's written by a successful entrepreneur, someone who has actually been through the start-up process from start to finish."

PETER G. - Amazon Reviewer

Project Management for Startups

Simple Tools and Templates to Launch, Lead, and Thrive

Murray Longe

WHITMAN
CROSS

Published by Whitman Cross Publishing, Melbourne, Australia.

First edition 2025
Second edition 2025
Third edition 2026

ISBN 978-1-7645437-3-6 (ebook)
ISBN 978-1-7645437-4-3 (paperback)
ISBN 978-1-7645437-5-0 (hardback)

NATIONAL LIBRARY OF AUSTRALIA

A catalogue record for this book is available from the National Library of Australia

Disclaimer

The information in this book represents the author's views as of the date of publication. It is for general informational purposes only and is not intended as professional, legal, financial, or business advice. Because circumstances and regulations change, the author reserves the right to revise his views in light of new conditions. Readers should consider their own personal circumstances and consult with qualified professionals before taking any action based on the information in this book. While every effort has been made to ensure the accuracy and completeness of the information, the author and publisher assume no responsibility for any errors, omissions, or inaccuracies. Use of the information is at the reader's own risk.

For more information, visit whitmancross.com

WHITMAN
CROSS

Also by Murray Longe

How To Start Your Start-Up:
The 10 Key Steps To Get From Idea To Investment

The Startup Success Series

#1 *Find Problems Worth Solving:*
How to Validate Your Startup Idea

#2 *Beyond The Minimum Viable Product:*
Build Products Customers Really Want

#3 *Scale Without Breaking:*
How to Grow Your Startup Without Losing What Made It Work

#4 *Beyond The Pitch Deck:*
How to Raise Startup Capital

Dedicated to everyone who has stood shoulder-to-shoulder with me on challenging projects.

Your support, teamwork, and resilience have turned even the toughest journeys into rewarding and enjoyable experiences.

Successful projects are never achieved alone – they are the result of shared commitment, creativity, and courage.

This book celebrates all of you.

Table of Contents

Table of Contents

Introduction ::
Taming Chaos Without Killing Creativity

If you've ever tried juggling flaming torches while riding a unicycle, you already know what running a startup or growing business feels like. It's thrilling, terrifying, and precariously balanced all at once. Now imagine doing it blindfolded. That's what managing a project without a plan looks like.

Welcome to the world of project management, startup style: equal parts chaos and possibility.

Emerging businesses—whether you call them startups, new ventures, or growing enterprises—thrive on creativity, speed, and making the impossible possible. They operate in fast-paced, high-pressure environments where uncertainty is a constant companion and resources are often stretched to the limit. Yet, amidst the chaos and creativity, one thing remains clear: a great idea will only go so far without the structure to bring it to life. That's where project management comes in. No, I don't mean endless meetings, bloated processes, or soul-draining bureaucracy. Good project management is simpler. It aligns your team, resources, and goals so you can actually build something extraordinary.

Let's quickly address a common misconception. Great project managers aren't just human calendars ticking off dates. Anyone with a calendar can do that. The truly valuable project managers - the ones who become indispensable to a growing business - anticipate challenges, navigate around roadblocks, and arrive ready with solutions in hand. They're proactive guides, not passive observers.

After four decades of managing projects, both massive and small, from global collaborations to scrappy startups and from partnerships across 150 organizations to teams of two people, I've seen it all - the spectacular successes, the painful failures, and everything in between.

And what I've learned is this: project management shouldn't create complexity; it should eliminate it.

Why Emerging Businesses Need Project Management (Even If They Think They Don't)

Emerging businesses operate in a reality that big corporations don't: tight budgets, tighter deadlines, and team members who are wearing too many hats. And while "move fast and break things" might sound cool on a motivational poster, in reality, breaking things without a plan just leaves you with a mess you don't have time to fix.

Let's be honest: traditional project management—with its heavy jargon (Agile, Scrum, Waterfall)—can feel like a bad fit for a fast growing business culture. These frameworks assume luxuries you likely don't have: endless resources, certified professionals, and plenty of time.

I hear it all the time: *"We're nimble - we don't need red tape slowing us down."*

They're right about the red tape - but wrong about the need for structure.

Good news: you can (and should) ditch the jargon and adopt lightweight, flexible project management that suits your culture and goals.

The Sweet Spot Between Chaos and Structure

Think of your business as a band. Too much structure kills improvisation. Too little, and it's noise. The sweet spot is that place where creativity and clarity come together. Without it, priorities shift, tasks slip through cracks, teams burn out, and momentum stalls.

With it, you pivot quickly, make confident decisions, and turn ideas into results. Unlike corporations weighed down by approvals and meetings, you can adopt just enough structure to free your creativity—not stifle it.

Project Management for Humans, Not Robots

At its core, great project management isn't about processes and templates—it's about people. Your team brings projects to life. Effective project management simplifies complexity and empowers everyone to do their best work. Over the years, I've shaped personal principles that guide every project I manage:

- **Keep it simple:** Complexity is the enemy of clarity. Complicated tools or processes kill momentum. Stick to clear, straightforward approaches.
- **People matter:** Your team makes or breaks projects—not your templates. Recognize their contributions, celebrate the wins (big and small), build trust, and actively support their growth. Great projects are always team efforts.
- **Positivity wins:** A positive, solutions-focused mindset turns obstacles into opportunities. When setbacks happen (and they will), optimism and creativity carry you forward.

- **Dates are sacred:** Realistic deadlines matter. Missed dates have ripple effects, impacting everything from revenue and marketing to stakeholder confidence. Once you set dates, your role is to protect them fiercely.
- **Watch the critical path:** Some tasks are make-or-break. Identify tasks that can derail your entire project if delayed, and give them extra attention. Stay alert—the critical path can shift as your project evolves.
- **Nurture your team:** Your relationship with each team member is your secret weapon and your team's wellbeing is directly linked to your project's health. Build genuine relationships, understand individual challenges, and stay attuned to team dynamics. Protect and nurture your greatest asset—your people.
- **Understand the stakes:** Your project isn't isolated—it's crucial to your business's growth, survival, and reputation. Understand exactly how your project impacts broader business goals, investor confidence, and stakeholder plans. Never lose sight of the big picture.
- **Communicate openly:** Transparent communication builds trust. Avoid surprises by regularly updating your team, stakeholders, and partners—good news and bad. People understand challenges arise; they value clarity about what you'll do next.
- **Constantly scan the horizon:** Exceptional project managers proactively look ahead. Develop the habit of asking, "What could go wrong next?" Spot potential roadblocks early and have contingency plans ready before you need them.

These principles are woven into the chapters ahead. They're designed to help you lead with clarity, confidence, and a healthy dose of common sense. To help you navigate uncertainty, inspire your team, and deliver results - whether you're launching a product, raising funding, or scaling your operations.

What You'll Get From This Book

This isn't a textbook filled with project management theory. We all know the world doesn't need another one of those. This is a practical guide born from four decades in the trenches.

You won't find theoretical fluff here. You will find a toolkit of simple, effective techniques you can start using today. We'll cover everything from setting clear goals to confidently managing a team (or yourself). We'll talk about real-world challenges - like what to do when your budget evaporates halfway through or when your dev and marketing teams speak completely different languages.

You'll get clear, actionable advice and you'll find tools you can adapt immediately to your own projects.

Inside, you'll find:

- **Clear, actionable advice:** From setting smart goals to managing unexpected setbacks.
- **Real-world examples:** Stories from startups that illustrate what works (and what doesn't).
- **Immediately useful tools:** Templates and techniques you can put into practice right away.
- **A human-centered mindset:** Empathy, clarity, and flexibility—because great project managers lead people, not just projects.

Whether you're managing your very first project or you're a seasoned leader refining your approach, this book meets you where you are and takes you further.

Your First Project Management Success Story Starts Here

You don't need expensive software, complicated certifications, or armies of assistants. You just need a willingness to learn, adapt, and take the first step.

Grab a coffee (or something stronger, depending on how your day's going) and let's get started. It's time to turn chaos into clarity, vision into results, and your big ideas into reality.

Your first project management success story is waiting to be written, and it starts right here.

1

Understanding Project Management for Startups and Small Businesses

1. What Is Project Management?

Turning ideas into reality

Project management is how ideas become reality. It's the discipline of planning, organizing, and guiding work to achieve specific goals - on time, on budget, and with impact.

Let's be honest, though: the term "project management" doesn't always inspire excitement. It might bring to mind Gantt charts, rigid checklists, and endless meetings. But forget all that for a moment. While those tools and practices might be part of it, project management at its core is much simpler - and far more powerful.

At its heart, **project management is the art and science of getting things done**. It involves turning ideas into reality through planning, organizing, and leading a team to achieve specific goals within a defined timeframe and budget. In emerging businesses, where resources are limited, time is valuable, and uncertainty prevails, project management isn't a luxury - it's a necessity.

If you've ever launched a product, run a marketing campaign, built a pitch deck, or organized a team effort - you've already been doing project management. The question is, how effectively?

This book is about helping you do it better - with more clarity, less chaos, and a lot more confidence. Done right, project management doesn't stifle creativity - it amplifies it.

Why Project Management Matters in Startups

In startups and small teams, where time is tight, budgets are small, and everything feels urgent, project management isn't optional - it's essential.

Without it, great ideas stall. To-do lists get buried. Deadlines slip. Energy gets wasted.

But with it? You create direction. You move faster with less stress. You align your team and give yourself the best shot at success.

Why it's critical in early-stage businesses:

- You're juggling multiple goals - often with limited help.
- You need focus and momentum, not bureaucracy.
- You must deliver results quickly while adapting on the fly.

Project management is your steering wheel. It keeps you pointed in the right direction, even when the terrain gets bumpy.

Breaking It Down: The Five Pillars of Every Project

At its core, project management consists of five key components. Nearly every project follows these five basic stages:

1. **Initiation**

   This is the moment your big idea is born. You figure out what you are trying to achieve, why it matters, and whether it's even possible. Clear goals are your starting point, whether you're launching a new product, scaling your operations, or meeting a critical milestone.

   - *Think of it as the brainstorming phase but with a dash of reality-check.*

2. **Planning**

   Here's where the magic happens - or at least, the spreadsheets. How will you achieve those goals? Here, you map out how you'll get from idea to outcome. You set goals, assign responsibilities, estimate costs, define timelines, and choose the tools you'll use. This is where you figure out what resources you need, set a timeline, and create a roadmap to get from point A to point B without taking a detour through point Z.

   - *It's where chaos becomes a clear plan.*

3. **Execution**

   This is the "roll-up-your-sleeves-and-get-to-work" phase. Tasks are assigned, things get built, tasks are done, meetings happen (hopefully sparingly), and progress is made. This is the action phase - getting the work done.

   - *This is where momentum happens - and where leadership counts.*

4. **Monitoring Progress & Adjusting**

   No project ever goes 100% according to plan (if yours does, I'd like to meet you and learn your secret). This phase is about tracking progress, measuring results, solving problems, and making adjustments as needed to stay on course. Monitoring ensures you're staying aligned with your goals and can pivot quickly if needed.

- *It's your chance to course-correct in real time.*

5. **Closure:**

   Ah, the sweet satisfaction of wrapping things up. This is when you deliver the final product, celebrate your hard work, and reflect on what went well - and what you'd do differently next time. And also take time to celebrate the win - even a small one.

   - *This is how you grow from one project to the next.*

This process might sound straightforward, but the way you approach it can make or break a project.

Why Should You Care?

If you're thinking, "That's nice, but I'm just trying to get my startup off the ground or keep my venture going without losing my mind", I hear you. But here's the truth: no matter how small your operation is, you're already doing some form of project management. Every time you set a goal, make a plan, or organize your to-do list, you're dabbling in the craft. The question isn't if you're doing project management - it's how well you're doing it.

Without a clear plan, even the most exciting ideas can fizzle out faster than a sparkler in a rainstorm. Tasks get forgotten, deadlines slip, and before you know it, you're stuck in a spiral of frustration and missed opportunities. Good project management doesn't just keep you on track - it helps you work smarter and avoid burnout.

Emerging Businesses Need Their Own Approach to Project Management

Traditional project management is often designed for large organizations with established processes, big budgets, approval chains, formal documentation, and plenty of time. Smaller ventures, on the other hand, operate in an entirely different reality:

- **Resources are limited:** Every dollar and every hour counts.
- **Timelines are tight:** Speed is often the difference between success and irrelevance.
- **Uncertainty is high:** Plans can change, priorities shift, and you must adapt quickly.

That's why emerging businesses need a project management approach that's **lean, flexible,** and **people-centered.** This approach focuses on what's essential, cuts out the noise, and empowers your team to deliver their best work.

Keep it Simple, Keep It Moving

One of the most important lessons I've learned in my career managing projects big and small is this: **complexity kills momentum.**

Complex tools and overly detailed plans might look impressive, but they often slow things down and obscure what truly matters.

In emerging businesses, where clarity and speed are everything, simplicity should be your guiding principle. Keep your processes lean, your tools straightforward, and your communication crystal clear.

Remember, the goal of project management isn't to create complexity - it's to cut through it.

The People Factor

Another truth about project management that is often overlooked is that **people are your greatest asset.** Tools and processes can only take you so far. Your team's creativity, effort, and collaboration bring projects to life.

Behind every successful project are people who make it happen.

As a founder or leader, your role isn't just to manage tasks - it's to lead people. That means:

- **Building trust:** Open communication and mutual respect create a foundation for success.
- **Recognizing contributions:** Celebrate the wins, big and small, to keep morale high.
- **Supporting growth:** Mistakes will happen. Use them as opportunities to learn and improve.

When you put people at the center of your project management approach, you don't just get better results - you build a culture of trust, collaboration, and shared ownership.

What Project Management Is Not

Before we dive deeper, let's bust a few myths about project management:

1. **It's not about micromanaging:**
    - Great project managers don't hover over every task. They empower their teams to take ownership and deliver results.

2. **It's not about rigid plans:**
    - Plans are important, but adaptability is essential. Be ready to pivot when the situation demands it.
3. **It's not about endless reporting:**
    - Stakeholders need updates, but overloading them with information wastes time. Keep reporting concise and useful.

The best project managers remove friction - they don't create it.

Shed these misconceptions, and you can embrace project management as a tool for empowerment, not a constraint.

Your Advantage

The good news is emerging businesses have an edge: unlike big corporations with bloated processes and too many cooks in the kitchen, you can move fast, pivot quickly, and experiment boldly. But speed without direction is just chaos. That's where a lightweight, startup-friendly approach to project management comes in.

You don't need to adopt some fancy, complicated methodology. What you need is a straightforward system that gives you clarity, keeps your team aligned, and makes your goals feel achievable instead of overwhelming. And that's exactly what this book will teach you.

A Simple Starting Point

So, where do you begin?

You start by recognizing that project management isn't something you'll "get to later." It's how you make progress - now.

It isn't the enemy - it's your secret weapon.

It doesn't matter if you're planning a product launch, designing a marketing campaign, or organizing a bake sale to fund your next round of development. If you want to get things done efficiently and effectively, project management is your best friend.

Think of it as your personal GPS - you might know your destination, but without a plan, you're just wandering. With the right tools and mindset, you can chart your course, navigate around roadblocks, and get where you're going faster - and with fewer headaches.

Let's dive deeper and get you building projects that don't just get finished - but make a real difference.

2. Why Traditional Approaches Don't Always Work

Over the years, I've worked with just about every type of project imaginable - from large-scale projects with multi-million-dollar budgets to early-stage startups fighting to get their first product out the door. And if there's one thing I've learned, it's this:

Traditional project management approaches rarely work well in startups.

They weren't designed for it. When most people hear "project management," they imagine environments with stable conditions, defined roles, and long planning cycles, rigid methodologies, endless paperwork, and complex tools designed to micromanage every task. These approaches can work well in large, established organizations where structure and predictability reign. But in an emerging business? They're more likely to cause frustration than results.

Startups operate in a completely different world - one where priorities shift overnight, resources are limited, and momentum is often the single greatest competitive advantage. A world of rapid change and constant experimentation. They thrive on flexibility and innovation, which traditional project management approaches often fail to accommodate. To succeed in this dynamic environment, emerging businesses need an approach that's lean, adaptable, and focused on outcomes rather than processes.

The Corporate Playbook Doesn't Fit

Let's get specific. Project management methodologies like Waterfall, PRINCE2, or even textbook Agile were created for large teams with layered decision-making and clear lines of accountability. They rely on detailed, upfront planning and linear execution, assuming that once the plan is made, it won't need to change. They assume you have a full-time project manager, dedicated QA staff, business analysts, and department heads.

In an emerging business, you might be all those people - and the person making the coffee.

Your team is small. Your budget is tight. And your timeline? Non-negotiable. While a corporation might spend three months finalising a project scope, you'll be lucky to get three days before something changes.

Emerging businesses operate in an entirely different reality:

1. **Uncertainty Rules:**
   In startups, priorities shift, market conditions change, and new opportunities emerge - often daily. Traditional approaches struggle to accommodate this level of unpredictability.

2. **Speed Is Critical:**

   Growing enterprises can't afford to spend weeks or months on detailed planning and documentation. They need to act quickly, learn fast, and adapt even faster.

3. **Limited Resources:**

   With small teams and tight budgets, emerging businesses can't support the administrative overhead that traditional project management often requires.

4. **Creativity Thrives on Freedom:**

   Overly rigid structures can stifle the creativity and innovation that startups rely on to differentiate themselves.

When founders and business owners try to implement traditional methods in this kind of environment, what they often end up with is overkill: too much process, not enough progress.

Example: Why Traditional Methods Fail in Startups

Imagine a startup developing a new app. Using a traditional approach, they'd gather requirements, create a detailed plan, and lock in deliverables before any work begins. But six weeks into development, user feedback reveals a critical flaw in the design. Pivoting now means scrapping most of the original plan, wasting precious time and resources.

In contrast, a leaner, more flexible approach would have prioritized rapid prototyping and iterative testing. The startup could adapt quickly without derailing the entire project by focusing on small, actionable goals and incorporating feedback early.

The Complexity Trap

One of the biggest challenges with traditional project management is its tendency toward complexity. From multi-layered reporting structures to overly detailed tools, complexity can overwhelm teams, slow progress, and obscure what truly matters.

I've seen it play out too many times: business owners trying to be "disciplined" end up buried under unnecessary documentation, redundant tools, and procedural overhead that adds zero value. Before they know it, their lean, energetic team is spending more time updating reports than building the actual product.

I've learned a simple truth: **complexity is the enemy of clarity and kills velocity.** Where every moment counts, you need systems and processes that support your goals, not bog you down. And without velocity, you're not going to survive long enough to worry about the perfect process.

Where Traditional Methods Fall Short in Startups

To put a finer point on it, here are some of the most common disconnects:

- **Complexity Overload:** Frameworks designed for scale often introduce too many moving parts. Startups need simplicity, not scaffolding.
- **Inflexibility:** Traditional methods assume a fixed plan. Startups change course constantly. You need systems that evolve with your business.
- **Resource Assumptions:** Corporate methodologies rely on specialist roles. In reality, your "team" might be a generalist founder and a freelancer working part-time from Bali.
- **Delayed Value:** Long planning cycles make sense when the business is big, the budget is large, and the cost of failure is high. In a startup, not launching fast enough *is* a failure.

The Project Manager as a Proactive Problem Solver

The most valuable project managers are much more than just 'date keepers' who mechanically track tasks and report on whether they were completed on time. Anyone with a calendar app can do that.

Project managers who make a real difference - the ones who become indispensable to a growing business - are those who can look 'over the horizon', anticipate problems before they materialize, and adjust course accordingly. They're the ones who don't just react to issues but proactively solve problems that threaten to derail the project. When challenges inevitably arise, they don't just report the bad news - they come armed with potential solutions, rallying the team to find creative workarounds while keeping stakeholders informed. In a startup or emerging business where resources are precious and every role must justify its existence, there's simply no room for 'coasters.'

The best project managers add tangible value by turning potential disasters into mere speed bumps through foresight, ingenuity, and collaborative problem-solving.

What Startups Actually Need

So what does effective project management look like in a startup?

It's not about ditching structure entirely - it's about applying just enough of it, in the right places, to help you move faster, not slower.

- **Clarity Without Bureaucracy:** A roadmap that keeps everyone aligned, without the layers of admin.
- **Adaptability by Design:** A system built to absorb change, not resist it.

- **Tools That Serve the Team:** Lightweight, intuitive tools that fit into how you already work - not tools that require training courses to operate.
- **Empowerment Over Oversight:** Systems that support autonomy and responsibility, not control for control's sake.

The Opportunity You Have

Startups have a huge advantage here: **you're not locked into legacy systems or old habits**. You can build a project management approach that suits your team, your culture, and the way you actually work.

You don't need a bloated framework designed for managing satellite deployments. You need a lean, flexible way to keep your startup focused, moving, and delivering. One that's designed for change, not stability. One that keeps your people engaged, your goals clear, and your progress visible.

That's what we're going to build in this book. Not theory. Not textbook Agile. Not process for its own sake. But a practical, founder-tested approach that fits the reality of startup life - and helps you turn ideas into outcomes, fast.

3. The Startup Advantage: Turning Constraints into Strengths

Over the years, I've managed projects at every scale - from large, multi-year initiatives with generous budgets, to lean, fast-moving ventures working out of shared spaces and running on optimism and coffee. And if there's one truth that spans the spectrum, it's this:

Constraints, when approached well, can become your greatest advantage.

Whether you're building a startup from scratch, scaling an emerging business, or steering a growing enterprise into its next phase, limitations aren't just part of the deal - they're often the spark that ignites creativity, focus, and agility.

Speed Over Scale

One of the defining advantages of a venture in its early or growth stages is speed. Large organisations, with their layers of approval and legacy systems, often resemble cargo ships: stable but painfully slow to change course. You, on the other hand, are captaining a speedboat.

In a growing enterprise, decisions can happen quickly. You're close to the action, close to your customers, and unburdened by bureaucracy. That agility gives you an edge - not just in execution, but in seizing fleeting opportunities that big players won't even see coming.

When you spot a shift in the market, you can respond in hours, not quarters. That responsiveness is a competitive advantage - one you should protect fiercely.

Constraints Fuel Innovation

It's easy to wish for more resources - more people, more time, more money. But having less can push you to be smarter.

Throughout my career, I've consistently seen the best ideas emerge from businesses with the least room to waste. A startup operating on limited funds often finds a breakthrough not because it had everything - but because it *didn't*.

- **Lean Budgets**
  Drive clarity. You're forced to identify what truly matters and deliver value with minimal waste.

- **Small Teams**
  Eliminate friction. Communication is direct, decisions happen fast, and there's a shared sense of purpose.

- **Tight Timelines**
  Prevent overengineering. Deadlines push you to produce something real, not endlessly polish the abstract.
- **Ongoing Uncertainty**
  Keeps you adaptive. Without rigid long-term plans, you're free to adjust course as conditions change.

In an emerging business, these constraints don't have to be obstacles - they can be design features of how you grow smarter, leaner, and faster.

The Power to Pivot

Larger companies often double down on failing projects, partly because they can absorb the cost - and partly because they can't admit the plan needs to change.

But if you're leading a venture, you don't have that luxury. And that's a good thing.

The ability to pivot quickly - based on feedback, data, or instinct - is one of the great superpowers of small and medium-sized enterprises. You're not locked into a single direction. You can shift, adapt, and reallocate resources swiftly, without dragging a bureaucracy behind you.

Slack did it. So have thousands of other businesses that found unexpected success not by staying the course, but by knowing when to change it. The same freedom exists in your growing enterprise - if you've built the courage and systems to act on it.

Focus Wins

One benefit of limited resources that rarely gets enough credit is the *gift of focus*. When you can't do everything, you're forced to choose what matters.

In larger corporations, it's easy to get lost in "nice-to-haves" and side projects that drain time and energy. But in a venture with a small, committed team, your constraints become a forcing function. You focus on high-impact work. You say no more often. You measure outcomes, not activity.

This discipline translates directly into sharper, faster project execution. You're not spread across ten initiatives - you're moving the needle on two that really count.

Building Systems for a Dynamic Environment

Let's not pretend this is all smooth sailing. Whether you're an early-stage startup or a growing enterprise on the cusp of scale, the reality is often messy.

Priorities shift. Plans unravel. What worked last quarter may suddenly stop delivering results. But this isn't a flaw in the system - it *is* the system.

What matters isn't removing the mess. It's building lightweight processes that help you navigate it. Not rigid frameworks that break under pressure, but flexible scaffolding that supports change, adapts fast, and keeps your people aligned.

You don't need control - you need clarity. You don't need red tape - you need rhythm.

Designing to Amplify Your Strengths

So how do you turn this environment into an operational advantage? You build your project management approach around the strengths you already have:

- **Speed**
  Choose fast, visual planning tools. Avoid long documents. Keep planning just-in-time and just-enough.
- **Agility**
  Work in short cycles. Review frequently. Encourage reflection and adaptation as a regular part of your rhythm.
- **Focus**
  Prioritise ruthlessly. Protect your team's time. Eliminate low-impact distractions that feel like progress but aren't.
- **Creativity**
  Create space for experimentation. Make failing fast an acceptable - and even expected - part of the process.

The Takeaway

Every venture - whether it's just getting started or already scaling - faces constraints. That's not the problem. The problem is how we respond to them.

When managed well, those limitations become accelerators. They sharpen your focus, drive your innovation, and help you build a culture that's capable of adapting faster than any corporate playbook ever could.

In the next chapter, we'll start translating these strengths into a practical, flexible system - one that works whether you're building from zero, navigating early traction, or scaling up a growing enterprise. Let's get to work.

2
Laying the Foundation for Success

1. Defining Your Project Objectives

Every successful project starts with a clear destination. If you don't know where you're going, any road might get you there - but it probably won't get you there efficiently, on budget, or in one piece. Without well-defined objectives, even the most talented team can get lost in a maze of competing priorities and unclear expectations. Where time and resources are limited, setting clear project objectives isn't just helpful - it's essential.

That's why clearly defining your project objectives is one of the most critical early moves you can make in any venture. Whether you're leading a fresh startup or a growing enterprise with a few years under its belt, well-crafted objectives turn good intentions into focused action. Your objectives serve as your project's North Star, guiding every decision, every task, and every conversation. They serve as your project's compass - helping your team stay aligned, measure progress, and make confident decisions in the face of inevitable change.

And yet, in the fast pace of an emerging business, it's tempting to skip this step and just *start*. I get it - you're moving fast, the pressure is on, and progress feels urgent. But skipping over objectives is like building a house without a blueprint. You'll stay busy, but you won't necessarily end up with what you set out to create.

Let's cut through the fluff and walk through a pragmatic, lightweight way to set objectives that *actually work.*

Scanning the Horizon While Setting Your Course

Setting objectives isn't just about defining endpoints - it's also establishing a framework for proactive management. Effective project managers use objectives as both their destination and their radar system, constantly scanning the horizon for obstacles that might appear between the present and that desired future state.

As you define your objectives, also consider, What potential problems might emerge on the way to achieving them? Which dependencies might shift? What assumptions are you making that carry risk? This forward-looking mindset transforms project planning from a one-time documentation exercise into an ongoing practice of anticipating challenges and preparing contingencies.

Remember, in startups, the path is rarely straight - the sooner you can spot potential detours, the more effectively you can navigate them without losing sight of your destination.

What Are Project Objectives, Really?

At their core, project objectives define what success looks like. They're not just aspirations or intentions - they're **concrete outcomes** your team is working toward. A well-formed objective answers three essential questions:

1. **What are we trying to accomplish?**
   Example: Launch a redesigned onboarding flow for new users.
2. **Why does this matter?**
   Example: To improve first-week retention and reduce early churn.
3. **How will we know if we've succeeded?**
   Example: Achieve a 20% increase in user retention within three months.

Good objectives eliminate ambiguity. They act as a unifying focus across departments, helping you avoid misalignment, wasted effort, and "busy work" that doesn't actually move the needle.

Why This Matters More in a Venture or Emerging Business

In early-stage ventures or leaner enterprises, **every hour and every dollar needs to count**. You don't have time for sprawling, unfocused projects or vague aspirations. Clear objectives give you:

- **Focus**
  You can't do everything, and you shouldn't try. Objectives help you focus on what truly matters.
- **Alignment**
  With clear goals, your team - no matter how small - can move in the same direction, with less handholding and fewer miscommunications.
- **Motivation**
  Tangible targets create shared purpose and forward momentum. It's easier to stay motivated when you know exactly what "success" looks like.

- **Accountability**
  Objectives become benchmarks. You can measure your progress and adjust quickly if you're off track.

How to Define Effective Objectives

Not all objectives are created equal. If they're vague or wishful, they'll drain your time and morale. Strong objectives follow the **SMART** structure:

1. **Specific**
   - Avoid vague goals. Clearly state what you're aiming for and what you want to achieve.
   - Example: "Increase website traffic by 30% over the next quarter" is better than "Get more visitors."
2. **Measurable**
   - What will you track to measure success? Metrics should be defined from the start.
   - Example: "Generate 500 qualified leads" is measurable. "Improve sales" is not.
3. **Achievable**
   - Be ambitious but realistic. Stretching Is good, while those that are impossible will demoralize your team and set yourself up to fail.
4. **Relevant**
   - Does this objective support your business's overall strategy and current priorities.
   - Example: If your goal is to enter a new market, your objective might be to build partnerships or acquire customers.
5. **Time-Bound**
   - Set a clear deadline to create urgency and accountability.
   - Example: "Someday" isn't a strategy, but "Deliver the project by October 31" sets a clear timeframe.

Use this framework as a filter. If your objective doesn't pass the SMART test, revise it until it does.

Real-World Example: Setting Objectives for a Product Launch

Let's say you're launching a new mobile app feature designed to improve user retention. Here's how you might define your objectives:

- **Specific:** Increase the retention rate of new users by 15%.

- **Measurable:** Track retention metrics using app analytics.
- **Achievable:** Based on past performance, a 15% increase is challenging but realistic.
- **Relevant:** Retention is a key driver of your startup's growth strategy.
- **Time-Bound:** Achieve the target within six months of the feature's launch.

With these objectives in place, your team knows exactly what they're working toward and why it matters.

The One-Page Project Objective

Don't overcomplicate it. You don't need a 15-page scope document that nobody reads. One page is enough to bring clarity and alignment.

A strong project objective statement should include:

- **The problem you're solving**
  Example: "Our customer onboarding is inconsistent, leading to low trial-to-paid conversions."
- **The desired outcome**
  Example: "Improve onboarding experience to increase conversion by 25% over the next quarter."
- **Key deliverables**
  Example: "Interactive onboarding sequence, revised email flow, updated knowledge base."
- **Success metrics**
  Example: "Track conversion rate, drop-off points, and time-to-first action."
- **Timeline**
  Example: "Deliver all components by July 15th."

Common Mistakes to Watch For

Even experienced teams can get this wrong. Here are a few traps to avoid:

- **Being too vague**
  "Improve customer experience" sounds noble but doesn't tell anyone what to do.
- **Overloading your list**
  Too many goals create confusion and dilute focus. Stick to one or two high-impact objectives at a time.
- **Setting unrealistic targets**
  Shooting for the moon is great - but not if your team gets disheartened before takeoff.

- **Skipping metrics**
  If you can't measure it, you won't know whether it's working - or when to change course.
- **Forgetting the "Why"**
  Objectives should always tie back to your broader mission and strategy.
- **Going solo**
  Objectives are stronger when defined collaboratively. Get input from your team early.

Bring Your Team Into the Process

In any growing enterprise or fast-paced venture, project objectives should never be imposed top-down. They work best when they're co-created.

- **Run a short workshop or brainstorm**
  Gather insights from those doing the work.
- **Tie objectives to roles**
  Make it clear how each person contributes to the broader outcome.
- **Refine collaboratively**
  Alignment breeds ownership. Ownership breeds accountability.

When people have a hand in shaping the goal, they're more likely to care about reaching it.

The Real Power of Objectives: Focus and Filters

One of the most valuable things a well-defined objective gives you is a **filter**.

When everything feels urgent (and it usually does in a startup or emerging business), your objective becomes a tool for saying no - to scope creep, to shiny distractions, and to good ideas that just aren't *right for now* ideas.

If something doesn't help you meet your objective, it's not the priority.

Looking Beyond the Immediate Goals

The most valuable project managers aren't just task-trackers - they're forward-thinking problem solvers.

While tracking milestones is part of the job, exceptional project managers constantly scan the horizon for potential issues before they materialize. They ask "What could derail us next week or next month?" rather than simply documenting what happened yesterday.

As you establish your role as a project manager, position yourself as this kind of value-adding problem anticipator, not merely a date keeper. In resource-constrained startups, everyone must contribute meaningful value - and for project managers, that value comes from preventing problems, not just reporting on them.

The Takeaway

Defining clear project objectives is the first step in turning your ideas into action. In a business where every resource counts and the stakes are high, objectives are your compass. They keep your team aligned, your efforts focused, and your progress measurable.

Before you move on, take 10 minutes to draft a one-page SMART objective for your next project. Share it with your team and see what they think.

You'll be amazed at how this single step can bring clarity to everything that follows.

In the next section, we'll take this objective and begin translating it into a simple, actionable project plan - a plan you and your team can actually use to make real progress, fast. Let's keep building your foundation for success!

2. The Startup-Friendly Project Charter

You've got a clear objective, a motivated team, and a promising project idea. Everyone's excited. The energy's high. But a few weeks in, something happens: deadlines slip, conversations get muddy, priorities shift, and no one's quite sure who's doing what anymore.

Sound familiar?

That's what happens when you launch into execution without alignment. And that's exactly where a **project charter** comes in - not as a bureaucratic burden, but as a sharp, simple tool for focus and clarity.

Let's reframe this right away: **a project charter isn't corporate red tape - it's your secret weapon for keeping your venture aligned and on track**.

A project charter is a high-level roadmap for your project. It defines what you're doing, why you're doing it, and how you're going to get there. A well-crafted project charter creates clarity and alignment without bogging you down in unnecessary details. Think of it as your project's elevator pitch: concise, purposeful, and focused on the essentials.

What Is a Project Charter (And Why Should You Care)?

A project charter is a concise, high-level document that answers the fundamental questions about your project:

1. **Why are we doing this?**
   What problem are we solving, or what opportunity are we chasing?
2. **What are we trying to achieve?**
   What are the specific, measurable outcomes we're aiming for?
3. **How will we make it happen?**
   Who's involved, what's the scope, and how will success be measured?

In practical terms, it's your project's **elevator pitch and game plan rolled into one**. It aligns your team, clarifies expectations, and serves as your reference point when things inevitably shift.

Importantly, this isn't a bloated 30-page document destined to die in a forgotten folder. This is **a living, working tool** tailored for ventures that need to move quickly, stay focused, and adapt with minimal friction. The charter is not a detailed project plan. Instead, it provides a high-level overview that keeps your team and stakeholders aligned.

Why Every Emerging Business Needs One

Whether you're a startup building your first MVP, or a growing enterprise launching your tenth product, a charter helps you avoid the most common execution pitfalls. In this environment, a project charter can:

- **Prevent Scope Creep:**
  A well-defined charter helps you avoid sneaky additions that dilute your project's focus and slow progress.
- **Create Alignment:**
  It ensures everyone - from founders to freelancers - understands what the project is, its purpose, goals and scope, why it matters, and where their responsibility lies from the outset.
- **Save Time:**
  By defining key parameters upfront, you can avoid misunderstandings and wasted effort later on.
- **Enable Quick Decisions:**
  With a clear charter in place, it's easier to evaluate whether a new idea or request aligns with the project's objectives.
- **Prevent Burnout and Frustration:**
  Nothing kills morale like ambiguity. A charter gives your team a structure they can rally behind.

What Goes Into a Lean, Impactful Project Charter

Forget the formalities and let's cut to what matters. A startup-friendly charter should fit on one or two pages. Here's what to include:

1. **Project Name:**
   Make it clear and memorable. A good name gives the initiative an identity.
   - A brief summary of what the project is about.
   - Example: "Project Phoenix: Developing a mobile app feature to improve user retention by 15% within six months."
2. **Objective Statement:**
   A short summary of what success looks like. Use the SMART format from the previous section.
   - What are the specific, measurable goals of the project?
   - Example: "Launch the feature by Q3 and achieve a 15% increase in retention among new users."
3. **Scope (In and Out):**
   Define the boundaries. What's included in this project - and just as important, what isn't?

- Define what's included - and excluded - in the project.
- Example: "This project focuses on feature development and user testing. Marketing campaigns and customer onboarding will be handled separately."

4. **Key Deliverables:**
   What are the tangible outputs of the project? Be specific.
5. **Timeline and Milestones:**
   Include start and end dates, plus a few key progress checkpoints.
   - High-level milestones and deadlines.
   - Example: "Prototype by Week 4, Beta Testing by Week 8, Launch by Week 12."
6. **Team and Roles:**
   Who's involved, and who's doing what? Clarity here avoids confusion later.
   - Identify who's involved, including team members, leaders, and external partners.
   - Example: "Product Manager (lead), Design Team, Engineering Team, and QA Specialist."
7. **Resources or Budget (if applicable):**
   Capture what's been committed to this project - time, money, people, tools.
   - A brief overview of the resources needed, including tools, funding, and team capacity.
   - Example: "$10,000 allocated for development tools and testing; engineering capacity at 20 hours per week."
8. **Success Metrics:**
   How will you know if it worked? Include 2-3 specific, measurable outcomes.
   - How will you know if the project is successful?
   - Example: "Retention among new users improves by 15%, as measured by app analytics."

Example: A Venture-Friendly Project Charter

Project Name:
"NovaLaunch - New Website for Product Expansion"

Objective Statement:
Redesign and launch our website to reduce bounce rate by 30% and increase lead conversion by 20% within 90 days.

Scope:
Included: Homepage, product pages, and core SEO improvements
Excluded: Blog content and support site (to be handled separately)

Key Deliverables:

- Final design mockups
- Live, mobile-optimized website
- On-page SEO implementation

Timeline:

- Kickoff: April 1
- Design complete: May 1
- Go live: June 30

Team and Roles:

- Jessie - Project Lead
- Maya - Developer
- Alex - UX Designer
- Quinn - SEO Specialist

Budget & Resources:

- $6,500 budget
- 25 hours/week across the core team

Success Metrics:

- Bounce rate reduced by 30%
- Average session duration increased by 15%
- 20% lift in lead capture via contact form

How to Build Yours - Quickly

You don't need to be a project manager to create a solid charter. Here's a simple process:

1. **Start with a short team meeting or working session**
   Align on goals, scope, and deliverables together.
2. **Collaboratively fill in the key sections**
   Use a shared doc or template - whatever works for your flow.
3. **Keep it tight**
   Aim for clarity over detail. Two pages, max.

4. **Share it with all stakeholders**
   Make sure everyone involved reads it, agrees with it, and refers to it.
5. **Use it**
   Don't let it die in a folder. Refer to the charter in check-ins, decisions, and retrospectives. It's your project's source of truth.

Why This Works

A well-designed charter becomes your **operational anchor**. When priorities shift, you can use it to ask: "Is this change in scope aligned with our original intent?" If not, you've got a clear basis to push back - or redefine direction intentionally, not reactively.

It's not about rigidity. It's about intentionality.

And the best part? Once you've done this a couple of times, you'll never want to run a project without one again.

The Takeaway

For any venture looking to deliver results - without the confusion, drama, or drift - **a lightweight project charter is non-negotiable**. It's not overhead. It's clarity, alignment, and confidence rolled into one sharp tool.

In the next section, we'll look at how to align your internal and external stakeholders, so the momentum you've created through your charter doesn't hit resistance when it matters most. Let's get everyone rowing in the same direction.

3. Stakeholder Alignment: Getting Everyone On Board

Even the best-laid project plans will stall if the people around you aren't moving in the same direction.

Imagine your venture as a rowing team - sharp, focused, moving at speed. Now imagine half the team is rowing backwards. That's what a project feels like without stakeholder alignment. You've got energy, drive, and good intentions - but progress turns to friction, and decisions get revisited over and over again.

Alignment isn't a luxury - it's a prerequisite.

Whether you're leading a lean startup, managing a cross-functional team in an emerging business, or scaling operations in a growing enterprise, you need the people around your project - co-founders, team members, clients, investors, advisors - to understand what's happening, why it matters, and how they fit in.

Why Stakeholder Alignment Matters

Stakeholder alignment means your entire support structure is working toward the same outcome. When alignment is missing, you see warning signs early:

- **Miscommunication** - Conflicting interpretations of what the project is actually about.
- **Conflicting Priorities** - One stakeholder wants speed, another wants more features. Neither gets what they want.
- **Wasted Time and Resources** - Decisions get revisited. Tasks are repeated. Frustration builds.
- **Loss of Momentum** - Team members lose interest, stakeholders disengage, and small cracks widen into real delays.

Alignment, done well, ensures that **clarity replaces chaos, and progress replaces politics**.

Who Are Your Stakeholders?

Stakeholders aren't just people with a title. They're anyone with influence, input, or interest in your project. They typically fall into two categories:

- **Internal Stakeholders**
  These are the people inside your venture - co-founders, team members, department leads - who are directly involved in or impacted by the project.

- **External Stakeholders**
  These are the clients, customers, investors, vendors, partners, and advisors whose input, approval, or collaboration is required for success.

Identifying and understanding the stakeholders in these groups is the first step toward alignment. Common stakeholders include:

- **Team Members:** The people directly working on the project.
- **Founders/Executives:** Leaders who set strategic priorities.
- **Investors/Advisors:** Those who provide funding and guidance.
- **Partners/Vendors:** External collaborators critical to the project's success.
- **Customers/End Users:** The ultimate beneficiaries of your project.

Each stakeholder group has unique needs and perspectives, and part of your job as a project manager is to balance these interests effectively. Not every stakeholder needs daily updates, but all the key players need to understand where you're going and how their involvement matters.

How to Achieve Alignment Without Bureaucracy

In a fast-moving environment, you don't have time for alignment theatre - lengthy meetings, overwrought decks, or endless consensus-building. Instead, focus on **clear communication, shared purpose, and lightweight structure**.

Here's how:

1. Start with a Kickoff That Sets the Tone

A well-run kickoff meeting is your chance to align expectations, establish ownership, and start strong. It doesn't need to be fancy - but it should be structured. Cover:

- **Why this project matters**
  Make the purpose clear and meaningful.
- **What success looks like**
  Connect back to the SMART objective and project charter.
- **Scope, timeline, and deliverables**
  Avoid surprises later by setting boundaries now.
- **Who's doing what**
  Clarify roles to prevent duplication, finger-pointing, or confusion.

Keep it interactive. Invite questions. Use this moment to generate shared ownership, not just nodding agreement.

2. Define Success Together

This is where many ventures go wrong: assuming everyone defines "success" the same way.

Don't assume. Ask.

- What outcomes matter most?
- What would "wildly successful" look like for each stakeholder?
- Are there any non-negotiables, constraints, or expectations that haven't been voiced?

By creating shared success criteria early, you avoid painful rewrites, backtracking, or friction later on. Everyone knows what they're building and why it matters.

3. Tailor Communication to the Audience

Different stakeholders care about different things. And they consume information in different ways.

- **Investors or advisors?**
  They want periodic, high-level updates that tie to strategy, milestones, and ROI.
- **Your team?**
  Needs clear, frequent updates focused on tasks, blockers, and deliverables.
- **Clients or partners?**
  Likely want visibility at key points - demos, check-ins, approvals - but not the day-to-day.

The right information, in the right format, at the right time, builds trust and keeps the wheels turning.

4. Use a RACI Chart to Clarify Roles

Especially helpful when you're juggling multiple contributors or cross-functional teams, a **RACI matrix** defines:

Role	Description
Responsible (R)	Person who does the work to complete the task
Accountable (A)	Person ultimately answerable for the task's completion
Consulted (C)	Person whose opinion is sought before decisions are made
Informed (I)	Person kept up-to-date on progress

It's simple, but it avoids the most common problem in project execution: **assumed ownership** that leads to dropped balls or duplicated effort.

5. Create Ongoing Feedback Loops

Alignment doesn't happen once. It's sustained.

Set a cadence that matches your project's pace:

- **Weekly or biweekly updates** - Quick, no-fluff progress reports.
- **Milestone check-ins** - Use natural breakpoints to pause, reflect, and realign.
- **Open channels** - Slack, email, or even a shared doc where stakeholders can ask questions or flag concerns anytime.

The key is to keep communication flowing - without overwhelming anyone.

When Alignment Goes Off Track

Even with the best setup, things can go sideways. Here's how to course-correct:

- **Conflicting priorities?**
  Return to your project charter. Re-anchor the conversation around the agreed objective.
- **Pushback from a key stakeholder?**
  Listen. Understand the concern. If needed, revisit the scope or timeline together.
- **Someone disengaged?**
  Reach out personally. Often, a one-on-one conversation surfaces issues that never make it into team meetings.

Don't ignore misalignment - it rarely fixes itself. Tackle it early and directly.

The Startup Superpower: Alignment Without Red Tape

In a growing enterprise or emerging business, alignment doesn't need to mean bureaucracy. It's not about having everyone agree all the time - it's about ensuring they understand what the project is, how they fit in, and why it matters.

The real magic? **Speed and clarity with just enough structure.** You're not trying to force uniformity. You're building shared direction.

The Takeaway

Stakeholder alignment is a multiplier. When you have it, everything moves faster, with fewer blockers and better outcomes. When you don't, even the best project plan can get bogged down in confusion and resistance.

By engaging stakeholders early, defining success collaboratively, tailoring communication, and creating simple feedback loops, you create the foundation for **a high-performing, high-trust project culture.**

In the next chapter, we'll shift gears and look at how to choose the right project approach - because the path you choose to get from idea to outcome depends on more than just best practices. It depends on *your* team, *your* context, and *your* goals. Let's dive into the options.

3

Choosing the Right Approach for Your Project

---

1. Methodology Matchmaking

Finding the Best Fit for Your Startup Project

Project management methodologies have a lot in common with dating apps. They all promise flexibility, results, and commitment to your success. But as anyone who's ever swiped left in frustration can tell you, compatibility matters more than the sales pitch.

In an emerging business or growing venture, choosing the right project approach isn't about picking the most popular method - it's about finding what works for *you*.

There's no one-size-fits-all approach, and you're not committing to a single approach forever - you're looking for the best fit for the project at hand. What's your project's complexity? How does your team operate? Are you iterating on an MVP or delivering a critical regulatory milestone? Your answers should shape your methodology - not the other way around.

This section is your shortcut through the jargon-filled maze of project management theory. No fluff. Just practical guidance to help you choose (or combine) a methodology that fits your goals, team, and timeline.

The Main Players: A Straight-Talking Overview

Let's break down the most common project management approaches and when they're useful in a fast-moving business environment.

1. Agile - The Adaptable Innovator

Agile thrives on iteration, continuous feedback, and progress in small, focused increments. It's a favorite among product and software teams for a reason: it handles uncertainty well and lets you pivot without blowing up the whole plan.

Best For:
Projects with changing requirements or high uncertainty - like product development, new features, or customer-driven innovation.

Strengths:

- Quick feedback loops
- Prioritizes value over process
- Encourages team autonomy

Watch Outs:

- Can feel chaotic without discipline
- Requires a team comfortable with ambiguity

In a Nutshell:
Ideal for ventures working on fast-evolving products, especially when user feedback drives decision-making.

2. Waterfall - The Methodical Planner

Waterfall is linear and structured: define everything up front, complete each stage before moving to the next. It's great when change isn't expected - or welcome.

Best For:
Projects with well-defined, fixed requirements - such as legal, compliance, or infrastructure work.

Strengths:

- Clear milestones and documentation
- Easy to manage for predictable, repeatable projects

Watch Outs:

- Inflexible once things are underway
- Doesn't adapt well to feedback or evolving needs

In a Nutshell:
Reliable and rigid. Best for parts of your growing enterprise that value stability over speed.

3. Lean - The Focused Minimalist

Lean focuses on efficiency: eliminate waste, streamline effort, and maximize value. It's not just a framework - it's a mindset.

Best For:
Emerging businesses focused on validating ideas quickly or improving internal processes.

Strengths:

- Laser focus on value
- Encourages experimentation and learning

Watch Outs:

- Can sacrifice quality if you move too fast
- Needs a deep understanding of customer value

In a Nutshell:
Perfect for MVPs, pilot programs, or trimming the fat from bloated internal processes.

4. Kanban - The Visual Flow Master

Kanban isn't a full-blown methodology - it's a visual workflow system. Tasks are organized into columns ("To Do," "In Progress," "Done") to make work visible and bottlenecks obvious.

Best For:
Ongoing workstreams like marketing, support, or maintenance tasks.

Strengths:

- Great for transparency
- Helps balance team capacity

Watch Outs:

- Doesn't provide long-term planning or prioritization
- Can get overloaded without discipline

In a Nutshell:
Ideal for teams managing continuous workflows across your venture.

5. Scrum - The High-Velocity Sprint Runner

Scrum is a flavor of Agile, designed for tight teams delivering fast. It divides work into time-boxed "sprints" (typically 1-4 weeks), with daily standups and clear roles like Product Owner and Scrum Master.

Best For:
Cross-functional teams working on short, focused delivery cycles.

Strengths:

- Strong team rhythm
- Frequent delivery of usable outcomes

Watch Outs:

- Can become process-heavy if not applied flexibly

- Requires experienced facilitation

In a Nutshell:
Excellent for ventures needing rapid prototyping and tight feedback loops.

6. Hybrid - The Custom Fit

Most small-to-medium ventures won't use just one methodology - and they shouldn't. Hybrid approaches combine the best parts of multiple systems to suit your unique context.

Best For:
Teams that need structure *and* adaptability. (So... most of us.)

Strengths:

- Tailored to your workflow
- Adapts as your business evolves

Watch Outs:

- Can get messy without clear boundaries
- Needs a strong understanding of the pieces you're mixing

In a Nutshell:
Startups and growing enterprises often find hybrid approaches are the sweet spot between speed, structure, and sanity.

How to Choose the Right Fit

Here are the key questions to help you find your match:

- **What's the nature of the project?**
  Is it fixed-scope and stable (Waterfall)? Or flexible and feedback-driven (Agile, Lean)?
- **How uncertain is the path forward?**
  If things are evolving rapidly, prioritize adaptability.
- **What's your timeline?**
  Need speed? Think Lean or Agile. Have time to plan and stage? Consider Waterfall elements.
- **What's your team like?**
  Are they process-oriented? Creative and autonomous? Match the system to their strengths.
- **How risk-tolerant are your stakeholders?**
  Agile thrives on iteration and learning. Waterfall offers predictability (on paper, at least).

Real-World Match: One Venture, Many Methods

Let's say you're leading an emerging business launching a new wellness platform. You might use:

1. **Product Development:**
   - **Approach:** Agile with Scrum.
   - **Reason:** Short sprints and regular releases. You're building an MVP and need to adapt features based on user feedback. Sprints allow you to iterate quickly.
2. **Marketing Campaign and Customer Support:**
   - **Approach:** Kanban.
   - **Reason:** Workflows that never stop. The marketing campaign involves ongoing tasks like content creation and ad optimization, which are best managed visually.
3. **Partnership Negotiations:**
   - **Approach:** Waterfall.
   - **Reason:** For legal and compliance deliverables, change isn't welcome. The process has a clear sequence: identifying partners, drafting contracts, and securing agreements.

Lean principles across the board to prioritize customer value and reduce waste.

That's a hybrid system, and it works precisely because it's built around the business - not a framework. By using different methodologies for different parts of the business, the startup ensures that each project runs smoothly without overcomplicating their processes.

Common Pitfalls to Avoid

1. **Forcing a Fit:**
   - Don't choose Agile just because it's trendy - or Waterfall because it's familiar or popular.
   - Stay flexible. Choose a methodology based on context. Tailor your approach to the project's needs, and don't be afraid to pivot.
2. **Overcomplicating the Simple:**
   - Not every project needs a full, complex framework, particularly for a straightforward project.
   - Keep it simple. Sometimes, a Kanban board and a daily check-in are enough. There's no need to overcomplicate it.

3. **Ignoring Your Team's Working Style:**
    - The best system is the one your team can actually use.
    - Involve them in the decision and choose and approach they're comfortable with.

The Takeaway

Choosing the right methodology isn't about being "Agile" or "Lean" or "Structured." It's about creating a system that **helps your team execute well** and **your project deliver real results.**

In ventures where change is constant and speed is essential, methodology is a means - not an end. Start simple. Stay adaptable. Mix and match as needed.

In the next section, we'll explore how to create your own custom approach - a practical, startup-tested blend that reflects your team's strengths, your project's demands, and your business's personality. Let's make it yours.

2. Mix and Match

Tailoring Your Approach

Mixing and matching project management methodologies allows you to strike the perfect balance between structure and flexibility. It's like crafting a playlist: you choose the right "tracks" to match the rhythm and mood of your project. The result? A tailored approach that's effective and distinctly your own, adaptable to your startup's unique challenges and opportunities.

And that's exactly the right instinct.

In a growing enterprise, a custom-fit approach isn't a luxury - it's a necessity. You move fast. You adapt constantly. Your team doesn't work like a corporate division - and that's a good thing. But it means rigid systems often fail to keep up with your reality.

This section is about building a project management style that fits *your* project, *your* people, and *your* goals. Because you don't need a perfect methodology - you need one that works.

Why Tailoring Your Approach Makes Sense

Let's get one thing clear: there's no one-size-fits-all approach in the real world. You're juggling fast timelines, evolving priorities, limited resources, and a team that wears multiple hats. A customized or hybrid methodology gives you the power to:

- **Adapt to your team**
  Different teams thrive under different rhythms. Your system should reflect how *your* people work best.
- **Scale sensibly**
  A two-person venture needs less process than a ten-person team - but your approach should be able to grow with you.
- **Cut through complexity**
  Focus on what moves the needle. Leave the rest behind.
- **Stay flexible under pressure**
  When priorities shift (and they will), you want a system that bends without breaking.

How to Build a Hybrid System That Actually Works

This doesn't need to be complicated. In fact, the goal is to keep it simple and effective. Here's how to do it:

Step 1: Start With Your Project's Needs

Ask the right questions upfront:

- What are we trying to achieve?
- Is this project exploratory or structured?
- How often are things likely to change?
- What's our timeline?
- What's our capacity - people, tools, budget?

Example:
If you're building an MVP with evolving requirements, you might lean into Agile sprints. If you're onboarding suppliers with clear legal steps, Waterfall might fit that phase better.

Step 2: Match the Method to Your Team

Your team's style matters. If they're disciplined and process-driven, some Waterfall or Scrum structure can work well. If they thrive on autonomy and creative flow, Lean or Kanban may serve you better.

Pro tip:
Start light. Add structure only where it adds value. If your project system starts to feel like a second job, it's too heavy.

Step 3: Choose Tools That Fit You - Not the Other Way Around

You don't need to adopt every tool that comes with a methodology. Choose tools that are intuitive, visible, and actually make things easier:

- **Kanban boards (e.g., Trello, Asana)** - Visualize progress and workflow.
- **Agile sprint tools (e.g., Jira, ClickUp)** - Manage short-term execution with tight feedback loops.
- **Simple shared docs or checklists** - For small teams, sometimes simple beats sophisticated.

Step 4: Borrow the Best: combine elements from the major methodologies

Methodology	Borrow This	Why It Works
Agile	Use sprints and regular check-ins	Keeps progress iterative and feedback-driven
Waterfall	Map high-level milestones and lock critical phases	Adds structure when parts of the project are fixed or dependent
Lean	Ruthlessly prioritize and cut waste	Keeps effort focused on value delivery
Kanban	Use visual boards and limit work-in-progress	Great for transparency and flow
Scrum	Define roles, run retrospectives	Builds discipline and team rhythm

Mix only what you need. Don't build a Frankenstein process - build something elegant and lean.

Step 5: Test, Learn, and Iterate

Your system should evolve with you. Set checkpoints to ask:

- Is this helping us work better?
- What's slowing us down?
- Where's the friction - tools, roles, clarity?

Don't treat your project system as sacred. Treat it as a prototype.

Real-World Examples

A Startup's MVP Launch

Let's say your venture is building a new mobile app MVP. Here's how your hybrid approach during the product build might look:

- **Planning phase (Waterfall):**
  Map out key milestones - user research, prototype, testing.
- **Execution phase (Agile):**
  Two-week sprints, weekly team stand-ups, focus on shipping features iteratively.
- **Tracking work (Kanban):**
  Board with columns like "To Do," "In Progress," and "Blocked" - for instant clarity.
- **Continuous feedback (Lean):**
  Constantly review user feedback and prioritize updates that deliver the most value.

This blend gives you *structure when you need it,* and *flexibility when things inevitably change*.

Launching A New Product Line

Imagine an emerging business is developing a subscription-based meal kit service. Here's how they might tailor their project management approach:

1. **Product Development:**
   - **Methodologies Used:** Agile + Scrum
   - **Why:** They're building an MVP and need to iterate based on customer feedback. Sprints keep the team focused on delivering small, incremental improvements.

2. **Marketing Campaign:**
   - **Methodologies Used:** Kanban
   - **Why:** The team needs to manage ongoing tasks like content creation and ad optimization without fixed deadlines.
3. **Supplier Onboarding:**
   - **Methodologies Used:** Waterfall
   - **Why:** Onboarding suppliers involves a clear sequence of steps, from contract negotiations to logistics setup.

By customizing their approach, the venture ensures each part of the business runs efficiently without overcomplicating processes.

Pitfalls to Avoid

1. **Overcomplicating the System**
   - **The Problem:** Combining too many elements can create confusion. If your "process" needs its own onboarding manual, you've gone too far.
   - **The Solution:** Stick to the essentials. Focus on a few key strengths that align with your project's needs.
2. **Ignoring Team Dynamics:**
   - **The Problem:** Using an approach that doesn't suit your team's workflow or preferences.
   - **The Solution:** Make sure your whole team is bought in. Involve your team in designing the process to ensure buy-in and usability. A great system no one uses is a waste.
3. **Lack of Adaptation:**
   - **The Problem:** Failing to adjust the approach when the project's needs change, or ignoring feedback.
   - **The Solution:** Check in with your team regularly. What's working? What's not? Then adjust.

The Takeaway

You don't need to follow someone else's system - you need to build your own. A hybrid approach gives you the freedom to blend structure and adaptability, discipline and creativity.

It's your toolkit. It should reflect how *you* work best - whether you're shipping your first feature or scaling your operations across teams.

In the next section, we'll talk about the role of intuition and experience - how gut feel, timing, and leadership instincts work *alongside* structured systems to drive smart, confident decision-making.

Let's keep building.

3. The Role of Intuition

Balancing Process with Instinct

No project plan survives contact with reality.

As any founder or project lead in a fast-moving venture can tell you, unexpected turns aren't the exception - they're the norm. You might have a carefully mapped-out sprint plan, a Kanban board humming along, or a roadmap that looks great on paper… but then something shifts: a customer says something you didn't expect, a team dynamic changes, or a competitor announces a surprise feature.

In that moment, it's not your process that gets you through - it's your *instinct*.

This is where experience, judgment, and intuition step in. And while intuition sometimes gets sidelined in business circles as "unscientific" or unreliable, in a growing enterprise where uncertainty is the norm, data is incomplete, and every decision feels high-stakes, blending process with intuition isn't just a nice-to-have - it's a survival skill.

What Intuition Really Is

Let's demystify it: intuition isn't magic. It's not guesswork either. It's your brain drawing on years of accumulated experience, knowledge, and pattern recognition. It's the internal compass you've built through wins, mistakes, reflection, and repetition.

When you say, "Something about this doesn't feel right," what you're often doing is accessing data - just not in spreadsheet form. That's *strategic intuition* at work.

And in high-ambiguity environments like startups, emerging businesses, or product-first ventures, this kind of fast, experience-informed thinking can be your best ally.

When to Trust Your Gut

There are moments in project delivery where your intuition should carry weight:

- **When data is incomplete or absent**
  In uncharted territory - new markets, novel features, early-stage launches - hard numbers are often scarce. Intuition helps fill the gaps.
- **When timing matters more than certainty**
  Overthinking can kill momentum. Sometimes, the best decision is the one you make quickly with conviction.

- **When reading people or situations**
  Hiring decisions, team morale, investor dynamics, and stakeholder alignment often come down to feel, not formulas.
- **When something just feels off**
  Maybe the plan looks solid, but your instinct says it's wrong. That's often the moment to slow down and reassess.

How to Balance Intuition With Process

Trusting your gut doesn't mean flying blind. The goal is to create a system where your instincts and your process *strengthen each other*. Here's how:

1. **Use Intuition for Direction, Process for Execution**

   Intuition is a powerful compass. Use it to decide *what* to do. Then let your systems and tools handle *how* to do it.

   Example:
   Your gut tells you to pivot from one product feature to another. Great - now your planning process helps the team shift cleanly and execute with clarity.

2. **Validate Instinct With Data**

   Your instincts are the signal. Data is the amplifier. Use customer feedback, usage metrics, or even small-scale testing to confirm (or challenge) your hunch.

   Gut: "Users aren't excited about this feature."
   Data: Run a quick survey or check engagement stats. Are they dropping off early? Are questions increasing?

3. **Build a Feedback Loop**

   The best way to strengthen your intuition is to reflect on how well it's served you. What worked? What didn't? Why?

   After-action reviews, team retrospectives, and personal reflection all help turn gut calls into conscious learning.

4. **Use Process as a Guardrail - Not a Straightjacket**

   Frameworks are there to support you, not to replace thinking. If a process feels wrong in a given moment, it probably is.

   Don't hesitate to adjust a sprint, skip a stand-up, or pull a team into an ad-hoc workshop if your instinct says it's the right move.

When to Be Cautious With Intuition

Even well-honed intuition can steer you wrong - especially under pressure. Watch out for these red flags:

- **You're emotionally over-invested**
  If ego or fear is driving the decision, your instincts may be distorted. Bring in a trusted voice to sanity-check your thinking.
- **The stakes are unusually high**
  Big-money bets or major strategic pivots need more than gut feel - run the numbers and pressure-test the logic.
- **You're reacting out of habit, not insight**
  Just because something worked before doesn't mean it will again. Look for new patterns, not just familiar ones.

Strengthening Your Intuition Over Time

Like any strategic tool, intuition can be sharpened with intention:

- **Review past decisions** - What did your gut tell you? How did it play out?
- **Expose yourself to new patterns** - Try new roles, new domains, or new industries. Stretch your mental database.
- **Collaborate and listen** - Instinct is enhanced by diverse perspectives. Stay curious about how others approach decisions.
- **Practice mindfulness** - Self-awareness helps you spot the difference between instinct and impulse.

Real-World Examples

Intuition in Action

Imagine you're leading a startup's product development team, and during a sprint review, you sense that the team isn't excited about the current direction. The metrics are fine, but your gut tells you something is off. Instead of sticking rigidly to the plan, you pause to gather feedback.

Your instincts were right: the team feels the current feature set isn't innovative enough to differentiate your product. Based on their input, you pivot the roadmap slightly, incorporating their ideas. The result? A more engaged team and a stronger product.

An Emerging Business in Decision Mode

Let's say your venture is preparing for a product launch. Everything looks good - testing went smoothly, marketing is queued, deadlines are in sight.

But you can't shake a feeling that the positioning isn't quite right. The value prop feels flat.

Instead of brushing it aside, you pause and test a few alternative messages with early customers. One variant significantly outperforms the original - and becomes the new messaging baseline.

That wasn't luck. That was intuition informed by experience - and validated by action.

Common Pitfalls and How to Avoid Them

1. **Over-Reliance on Intuition:**
   - **The Problem:** Making decisions based solely on gut feelings can lead to bias or oversight.
   - **The Solution:** Use intuition as a guide but validate it with data and team input.
2. **Ignoring Intuition Completely:**
   - **The Problem:** Blindly following processes without considering your instincts can result in missed opportunities.
   - **The Solution:** Trust your gut, especially when processes don't account for unique challenges.
3. **Confusing Fear with Intuition:**
   - **The Problem:** Fear of failure or change can masquerade as instinct, leading to overly cautious decisions.
   - **The Solution:** Distinguish between instinct-driven confidence and fear-driven hesitation.

The Takeaway

In the real world, especially in fast-paced, resource-constrained environments like startups and growing enterprises, **the most successful project leaders don't choose between process and instinct. They master both.**

Process keeps you grounded. Intuition helps you fly.

You're building systems to help your team move with clarity and consistency. But you're also building your judgment - so that when those moments of ambiguity arise, you're ready to move with confidence.

In the next chapter, we'll explore the tools and techniques that bring these systems to life - so you can simplify complexity, keep your team aligned, and move from idea to execution with less friction and more momentum.

Let's go make project management *actually work* for your venture.

4

Essential Tools and Templates

1. Tools That Don't Break the Bank

Affordable and Effective Solutions for Startups

When you're running a growing business, every dollar has a job. The idea of shelling out for high-end enterprise software - before you've hit product-market fit or hired your fifth team member - feels like madness. Fortunately, you don't have to.

Today's digital landscape is packed with affordable (and often free) tools designed with emerging businesses and growing ventures in mind. These platforms are lean, powerful, and built to scale with you - not drown you in unnecessary complexity or cost. They're designed to help you focus on what matters most: delivering results.

In this section, we'll explore tools that help you manage projects, communicate clearly, and stay on top of timelines - without burning your runway.

Using Tools to Anticipate Issues, Not Just Track Them

The tools you choose should do more than just document what's already happened - they should help you see what's coming.

The most effective project managers configure their tools as early warning systems, not just progress trackers. When evaluating any tool, ask: "Will this help me spot problems before they become critical?" Look for features that highlight dependencies, flag approaching deadlines, or visualize resource constraints before they impact delivery. For example, a simple color-coding system in your task board that shows items approaching their deadline can be more valuable than elaborate reporting features.

Remember, in resource-constrained environments, the ability to identify and address issues early is often the difference between project success and failure. Your tools should amplify your natural problem-solving instincts, not just create a digital record of past performance.

What Makes a Tool Worth Using

Let's set the criteria. A great venture-ready project management tool should be:

- **Affordable**
  Free, or cost-effective with scalable pricing as your team grows.
- **Intuitive**
  You shouldn't need a training course or onboarding webinar to use it.
- **Collaborative**
  Whether your team is co-located or fully remote, the tool should make teamwork easier.
- **Flexible**
  Startup priorities shift fast - your tools need to keep up.
- **Integratable**
  The best tools play nicely with your other platforms (email, calendars, file sharing, etc.).

The Core Stack: Categories and Recommendations

Here's your no-nonsense, founder-approved toolkit. Each tool has been chosen for its balance of value, usability, and startup-friendliness.

1. Task & Project Management Tools

1.1. Trello

https://trello.com

A visual task management tool based on Kanban-style boards for managing tasks with drag-and-drop simplicity. Great for organizing tasks, tracking progress, and collaborating with your team.

Best for: Visual thinkers and early-stage teams managing projects with clear, discrete tasks.

Pricing: Free tier available; paid plans from $5/user/month.

1.2. Asana

https://asana.com

A robust project management platform offering lists, timelines, boards, and automation. Offers more structure than Trello for complex projects.

Best for: Teams managing multiple projects with cross-functional collaboration across teams.

Pricing: Free tier available; paid plans from $10.99/user/month.

1.3. Notion

https://www.notion.so

A flexible all-in-one workspace for managing tasks, documents, wikis, and project dashboards.

Best for: Teams wanting structure *and* creativity in one space. A customizable tool for project management and knowledge sharing.
Pricing: Free tier available; paid plans from $8/user/month.

1.4. Microsoft Excel

https://www.microsoft.com/microsoft-365/excel
While often overlooked as a project management tool, Excel remains powerful, flexible, and familiar for managing projects of all sizes.
Best for: Teams who need customization options and are already familiar with Excel. Great for budget tracking, resource allocation, and creating Gantt charts without a learning curve.
Pricing: Included with Microsoft 365 (plans from $6/user/month); one-time purchase options are also available.

Why Excel Shouldn't Be Overlooked for Project Management

Despite the proliferation of dedicated project management software, Microsoft Excel continues to be a viable option for managing projects, especially for small to medium-sized initiatives. Here's why:

- **Universal familiarity**: Most professionals already know how to use Excel, eliminating the learning curve that comes with specialized PM tools. Everyone recognizes the grid structure of rows and columns, making it immediately accessible.
- **Flexibility and customization**: Unlike pre-defined structures in dedicated tools, Excel allows you to design your own columns and rows, apply formatting to track exactly what you need, and adapt quickly to changes in project scope or timeline.
- **Cost-effectiveness**: Excel is likely already part of your Microsoft 365 subscription, meaning there's no additional cost for a separate project management tool.
- **Powerful data analysis**: Excel's extensive library of formulas and functions lets you perform complex calculations, track project costs, analyze resource utilization, and create professional reports with conditional formatting.
- **Versatility**: For teams juggling multiple types of projects, Excel's adaptability allows you to customize your approach for each initiative rather than forcing everything into a one-size-fits-all solution.
- **Collaboration features**: Modern Excel supports real-time collaboration through OneDrive or Google Sheets, making it viable for team coordination.

2. Team Communication & Collaboration Tools

2.1. Slack

https://slack.com

The gold standard for team messaging. Organize discussions into channels, share files, and integrate with your other tools.

Best for: Teams that need real-time communication and easy integration with other tools replacing internal email and keeping communication quick and organized.

Pricing: Free tier available; paid plans from $7.25/user/month.

2.2. Google Chat (via Workspace)

https://workspace.google.com/products/chat

Simple, built-in team chat tool for businesses already using Gmail and Google Calendar.

Best for: Teams that need reliable, cloud-based collaboration tools wanting lightweight comms within Google Workspace.

Pricing: Included with Google Workspace (plans from $6/user/month).

2.3. Zoom

https://zoom.us

A trusted video conferencing platform for virtual meetings, team meetings, webinars, or customer demos.

Best for: Remote or hybrid teams that value reliable video communication.

Pricing: Free tier available; paid plans from $14.99/month.

3. Time Tracking & Deadline Management

3.1. Clockify

https://clockify.me

Track hours by project, task, or client. Great for freelancers and small teams to monitor how they're spending their time..

Best for: Ventures that want to improve productivity and track billable hours.

Pricing: Free tier available; paid plans from $3.99/user/month.

3.2. Toggl Track

https://toggl.com/track

User-friendly time tracking with detailed reports and integrations.

Best for: Teams juggling multiple deliverables with varied time requirements and billing clients.

Pricing: Free tier available; paid plans from $10/user/month.

3.3. Google Calendar

https://calendar.google.com

Set project deadlines, schedule recurring standups, and sync across time zones.

Best for: Lightweight scheduling that works across the team.
Pricing: Free; included in Google Workspace.

4. File Sharing & Collaboration

4.1. Google Drive

https://drive.google.com
Real-time document collaboration, version control, and easy file sharing.
Best for: Teams already using Google Workspace tool (Google Docs, Sheets, and Slides).
Pricing: Free for up to 15 GB; paid plans start at $1.99/month.

4.2. Dropbox

https://dropbox.com
Secure cloud storage with powerful file-sharing and collaboration capabilities.
Best for: Teams that need to share and access large files securely, like videos or design assets.
Pricing: Free for 2 GB; paid plans from $9.99/month.

5. Reporting, Analytics & Dashboards

5.1. Google Analytics

https://analytics.google.com
Track website traffic, user behavior, and funnel performance.
Best for: Businesses that need to measure the impact of online activities.
Pricing: Free.

5.2. Google Looker Studio (formerly Data Studio)

https://lookerstudio.google.com
A free tool to build interactive, shareable dashboards and visualize key metrics.
Best for: Turning spreadsheets and analytics into clear, shareable insights. Visualizing project or marketing data.
Pricing: Free.

5.3. Airtable

https://airtable.com
Part spreadsheet, part database, part project dashboard tool for tracking and analyzing project data.
Best for: Teams managing complex data or workflows in a customizable format.
Pricing: Free tier available; paid plans from $10/user/month.

How to Build Your Tool Stack Wisely

Avoid tool fatigue by taking a practical approach to adoption:

- **Start small.** Begin with one or two tools that solve immediate pain points.
- **Trial before you buy.** Use free versions and trials to get team feedback.
- **Ensure integration.** Choose tools that connect well with others - this saves time and avoids rework.
- **Review regularly.** Set a quarterly check-in to ask, "Is this still helping us move faster?"

Create Your "Startup Stack"

Here's a starter bundle you can build on:

Need	Tool Suggestion
Task Management	Trello, Asana, or Excel
Team Communication	Slack
File Sharing	Google Drive
Time Tracking	Toggl or Clockify
Scheduling	Google Calendar
Reporting	Looker Studio, Airtable, or Excel

This kind of curated stack delivers outsized value with minimal complexity.

The Takeaway

Managing projects on a budget doesn't mean compromising on quality. You don't need expensive software or heavyweight platforms to run a high-performing team. The best tools for your emerging business are the ones that are simple, affordable, and powerful enough to grow with you.

Your goal isn't to create the perfect tech stack - it's to create just enough structure to execute well, communicate clearly, and adapt quickly. These tools help you do exactly that.

In the next section, we'll explore templates - the time-saving shortcuts that help you build momentum from day one. From one-page charters to sprint plans, you'll find ready-made frameworks you can steal, adapt, and deploy immediately. Let's keep building your toolkit for success!

2. Templates You'll Actually Use

Streamline Your Work with Practical, Ready-to-Go Solutions

Let's be honest: most templates out there were built for corporate teams with too much time and not enough urgency. They're bloated, confusing, or so abstract you'd need a business analyst just to fill in the first line.

But when done right, templates can become one of your greatest productivity hacks. They simplify planning, reduce decision fatigue, and make collaboration smoother - especially when you're moving fast and wearing multiple hats.

This section gives you a library of startup-ready templates - lean, practical, and actually usable. These aren't forms to check boxes. They're tools to help your team stay focused, execute faster, and make better decisions under pressure.

Why Templates Matter

Templates are like shortcuts - they give you a head start without sacrificing quality. In a growing enterprise, your most limited resources are usually time, money, and mental bandwidth. Templates give you leverage. They:

- **Save time** - No more reinventing the wheel every time you launch a project or run a meeting.
- **Reduce friction** - Clear structures help your team get aligned faster.
- **Prevent mistakes** - Built-in reminders mean you don't miss crucial steps.
- **Keep everyone on the same page** - Templates create shared language and expectations.

Templates That Work in the Real World

Below are seven field-tested templates designed specifically for startups, founders, and growing teams.

1. The One-Page Project Plan

Template Link: Atlassian Project Plan
https://tinyurl.com/3fuc5xb2

Alternative: Sample Template
https://tinyurl.com/5xe9bn8f

Best for: Kicking off projects clearly and concisely.

What it covers:

- Project name & objective (SMART format)
- Problem being solved

- Key deliverables
- Scope (what's included/excluded)
- Milestones & timeline
- Team roles and responsibilities

When to use it: At the beginning of any project. Keep it short, simple, and visible.

2. Risk Management Template

Template Link: Sample Template
https://tinyurl.com/5xe9bn8f

Best for: Planning for the things that *might* go wrong - before they do.

Fields:

- Risk description
- Likelihood (High/Med/Low)
- Impact (High/Med/Low)
- Mitigation strategy
- Owner

When to use it: For high-stakes, complex, or multi-stakeholder projects.

3. Meeting Agenda & Notes Template

Template Link: Sample Template
https://tinyurl.com/5xe9bn8f

Best for: Running structured, outcome-focused meetings.

Covers:

- Meeting purpose
- Key agenda items
- Discussion points
- Action items
- Owners and due dates
- Notes / Decisions

When to use it: For team standups, stakeholder check-ins, or retros.

4. Post-Project Review (Retrospective) Template

Template Link: Sample Template
https://tinyurl.com/5xe9bn8f

Best for: Capturing lessons and building continuous improvement into your culture.

Sections:

- What went well
- What didn't
- Surprises
- Lessons learned
- What we'll do differently next time
- Follow-up actions

When to use it: After sprints, launches, or key deliverables.

5. Budget Tracker Template

Template Link: Excel Template
https://tinyurl.com/9xtbrrb8

Best for: Staying in control of project or campaign budgets.

Columns:

- Category (e.g., tools, talent, marketing)
- Budgeted amount
- Actual spend
- Variance
- Notes

When to use it: On any project with a financial component.

6. Kanban Board & Task Tracker Templates

Template Link: Trello's Kanban Template
https://tinyurl.com/2p26uvyw

Alternative: Notion's Board View Template
https://tinyurl.com/5dvdcjj

Alternative: Sample Template
https://tinyurl.com/5xe9bn8f

Task Trackers vs. Kanban Boards: Understanding the Differences

You may hear of tools Kanban Boards and Task Trackers being used interchangeably. While both tools help teams organize work, Task Trackers and Kanban Boards serve different purposes and operate on different principles.

Key Differences

Task Trackers are list-based tools for managing individual tasks:

- Focus on task status and completion
- Structured as lists with checkboxes
- Include metadata (due dates, assignees, priorities)
- Examples: Todoist, Microsoft To Do

Kanban Boards are workflow visualization tools:

- Focus on work moving through stages
- Represent tasks as cards across columns
- Highlight bottlenecks and work in progress
- Implement WIP limits
- Examples: Trello, physical boards, Kanban views in Asana/Jira

Simplifying with a Hybrid Approach

For growing businesses seeking simplicity, a single tool can effectively serve both purposes. You don't need two different tools! You can use one tool both ways:

A Kanban Board can be your task list if you put all the important details on each card.

A Task List can work like a Kanban Board if you add columns like "Not Started," "Working On It," and "Finished."

Choose a flexible tool offering both list and board views (Asana, ClickUp, Notion) to toggle between perspectives without duplicating systems.

The goal is to reduce overhead while maintaining visibility and control with an approach that fits your team's style.

Excel Project Management Templates

While the above templates work great in various tools, Excel offers a powerful yet familiar environment for project management. These Excel-specific templates can help you leverage a tool you likely already know to manage projects effectively.

7. Project Charter Template (Excel)

Template Link: Smartsheet Project Charter Template
https://tinyurl.com/3kpysdvr

Alternative: Continuous Improvement Toolkit Project Charter
https://tinyurl.com/5hf82hb8

Best for: Defining project scope and objectives clearly at the start.

Features:

- Comprehensive project overview
- Key stakeholder information
- Core objectives and deliverables
- Project assumptions and constraints
- Success criteria

Why use Excel for this: The familiar format makes it easy to share with stakeholders who may not use specialized PM tools.

8. Gantt Chart Template (Excel)

Template Link: Microsoft's Built-in Gantt Chart Templates
https://tinyurl.com/yc5kev4b

Best for: Visualizing your project timeline and milestones.

Features:

- Task dependencies and relationships
- Visual timeline of the entire project
- Milestone tracking
- Resource allocation
- Progress indicators

Why use Excel for this: Excel's grid structure and conditional formatting capabilities make it surprisingly effective for creating and managing Gantt charts without specialized software.

9. Resource Planning Template (Excel)

Template Link: Smartsheet Resource Planning Template
https://tinyurl.com/ya75bbje

Best for: Balancing team workloads and availability.

Features:

- Team member allocation across projects
- Skill tracking
- Availability management
- Workload balancing
- Resource forecasting

Why use Excel for this: Powerful formulas allow you to calculate utilization rates and identify overallocation issues quickly.

10. Project Dashboard Template (Excel)

Template Link: Smartsheet Project Management Dashboard https://tinyurl.com/ycxnd3sf

Best for: Visualizing project data for better decisions.

Features:

- KPI tracking
- Visual metrics using charts and graphs
- Status indicators
- Budget vs. actual comparisons
- Risk tracking

Why use Excel for this: Excel's robust charting capabilities make creating compelling visualizations straightforward for people who already know the tool.

11. Issue Tracker Template (Excel)

Template Link: Smartsheet Issue Tracker Template https://tinyurl.com/mr4dvrbt

Best for: Logging and addressing project issues promptly.

Features:

- Issue categorization
- Priority settings
- Status tracking
- Assignment to team members
- Resolution documentation

Why use Excel for this: Conditional formatting can easily highlight high-priority issues and approaching deadlines.

Why Consider Excel Templates for Project Management

While dedicated project management tools have their merits, Excel-based templates offer distinct advantages:

1. **Low learning curve**: Most team members already know how to use Excel, eliminating the barrier to adoption.

2. **Customization**: Unlike rigid PM software, Excel lets you modify templates to fit your exact needs and workflow.
3. **Integration**: Excel works seamlessly with the Microsoft ecosystem many businesses already use.
4. **Data analysis**: Excel's powerful formulas and functions allow for complex calculations and reporting that some dedicated PM tools lack.
5. **Accessibility**: Excel's familiar interface makes project information accessible to stakeholders who may be reluctant to learn new software.
6. **Cost-effectiveness**: You're likely already paying for Excel as part of Microsoft 365, avoiding additional software costs.

Where to Store and Share Templates

To keep things simple:

- **Use Google Drive** for documents and spreadsheets
- **Set up a Notion workspace** with linked templates and SOPs
- **Create a "Templates" channel in Slack** to drop links and updates

Pro Tip: Organize your templates by function - Planning, Execution, Meetings, Finance, and so on. This makes reuse and onboarding a breeze.

Where to Find More (Free) Templates

Here are a few excellent sources for pre-built, customizable templates:

- **Notion Startup Templates** - https://www.notion.so/templates
- **Trello Template Gallery** - https://trello.com/templates
- **Canva Project Plan Templates** - https://www.canva.com/templates/?query=project-planner
- **Google Workspace Templates** - https://docs.google.com/templates
- **Airtable Startup Templates** - https://www.airtable.com/templates
- **Microsoft Excel Templates** - https://templates.office.com/en-us/templates-for-excel

The Takeaway

The right template isn't a bureaucratic form - it's a time-saving, stress-reducing tool that helps your team move with clarity and confidence. Templates give you structure without locking you in. They're adaptable, repeatable, and - once you've found your rhythm - remarkably freeing.

Remember that familiar tools like Excel should not be overlooked simply because newer, specialized options exist. For many teams, the best solution is the one they already know how to use effectively. Excel offers powerful project management capabilities with virtually no learning curve, making it an excellent choice for teams who want to get started quickly without introducing new tools.

By using startup-friendly templates like these, you'll spend less time organizing and more time executing.

In the next section, we'll zoom out and talk about digital versus analog approaches. Because sometimes the best system isn't another app but a sticky note, a whiteboard, or a notebook you actually use. Let's keep building your toolkit with what works, not just what's shiny.

3. Digital vs. Analog: Finding the Right Balance

Leveraging Both Worlds for Startup Success

These days, it feels like every new productivity app promises to solve all your problems - automatically, seamlessly, and with a sleek UI that somehow knows your schedule better than you do. But if you've ever scribbled an idea on a napkin or mapped out a product launch on a whiteboard, you know that *sometimes* the simplest tools are the most powerful.

In project management, it's not about choosing digital *or* analog - it's about choosing what works. In this section, we'll see how to blend tech and tactile tools in a way that plays to your team's strengths, keeps things moving, and doesn't overcomplicate what should be simple.

Why Digital Tools Dominate (and Rightfully So)

There's no denying it: digital tools have become the backbone of modern startup operations. They're powerful, scalable, and - when used well - can save hours of manual coordination.

Here's why digital is often the default:

- **Collaboration Across Locations**
  Cloud-based platforms like Trello, Asana, and Slack allow teams to work in sync - whether they're across town or across time zones.
- **Automation = Time Saved**
  With tools like Zapier or built-in integrations, you can auto-assign tasks, schedule recurring check-ins, and track milestones with zero manual intervention.
- **Scalability Without Growing Pains**
  Start with two team members. Scale to 20. Your digital tools grow with you - no need to redraw a wall-sized Gantt chart.
- **Instant Search and Documentation**
  Need to find the project brief from three weeks ago? Just search your project hub. Done in seconds.
- **Integrations That Just Work**
  From Google Workspace to Notion, modern tools talk to each other. It's like building your own connected command centre.

When Analog Wins: Don't Ditch Your Whiteboard Just Yet

Despite all the software in the world, analog tools are still deeply valuable - especially when you need to think creatively, get focused, or just *get going* without a login.

Here's what analog brings to the table:

- When you're offline, there are no alerts, badges, or Slack pings to pull you away from what matters.

When to Use Which: A Quick Startup Guide

Situation	Best Fit
Brainstorming new ideas	Analog (whiteboard, sketchpad, sticky notes)
Cross-functional execution	Digital (Asana, Trello, Notion)
Team stand-ups (in-office)	Analog (wallboard or Kanban wall)
Remote collaboration	Digital (Slack, Google Docs, Zoom)
Project retrospectives	Either (Notion doc or sticky notes on a wall)
Personal daily planning	Analog (notebook, planner)
Tracking OKRs / KPIs	Digital (Sheets, dashboards, Airtable)

The Hybrid Approach: The Real Startup Secret

Most growing ventures use *both* - they just don't always realise it. The trick is being *intentional* about how you combine them.

Here's how to make that mix work:

1. **Plan Analog, Execute Digital**

   Use a whiteboard, paper sketch, or sticky notes for ideation and planning. Then, transfer refined plans to a digital workspace like Trello or ClickUp for tracking and ownership.

2. **Use Analog for Focus, Digital for Collaboration**

   Jot your personal priorities in a notebook. Use a shared task board for the team. This avoids clutter while keeping everyone aligned.

3. **Keep a Visual Workspace**

   Maintain a physical Kanban or progress board in the office for visibility - even if the source of truth lives in Notion or Asana.

4. Sync After Meetings

Ran a retro or brainstorm with sticky notes? Snap a photo and summarise key takeaways in Google Docs or Miro, where everyone can comment and reference later.

5. Personalise Your Stack

Designers may prefer sketchbooks. Ops leaders may lean into Gantt charts. Let your team choose the tools that suit *their* role - just make sure the core project updates flow back to one central source.

Excel: The Bridge Between Digital and Analog

Microsoft Excel occupies a unique middle ground in the digital-analog spectrum. While fully digital, its flexibility allows it to mirror the simplicity and customization of analog tools.

For teams transitioning from whiteboards and sticky notes, Excel provides:

- The familiar grid layout that resembles paper planning sheets
- The ability to customize and adapt formats without coding knowledge
- Visual elements like conditional formatting that provide at-a-glance status, similar to color-coded sticky notes
- Offline functionality when needed, unlike many cloud-only tools

Many successful startups begin with Excel-based project management and gradually transition to specialized tools only when they outgrow its capabilities – often much later than you might expect.

Avoiding Common Pitfalls

Pitfall	Solution
Duplicate Work	Don't track the same task in five places. Choose one source of truth for execution, and sync analog planning *into* that system.
Too Many Tools	Limit your toolkit. If you're constantly bouncing between apps, it's time to simplify.
Missed Sync Points	Document insights from analog sessions so your remote teammates aren't left out.

Real-World Example: Hybrid in Action

Scenario: Your venture is preparing for a product launch.

- **Kickoff Meeting (Analog)** - The team meets in-person to brainstorm on a whiteboard, mapping out objectives, milestones, and risks. You map out launch goals and a rough timeline using a whiteboard and sticky notes.

- **Task Assignment & Management (Digital)** - The agreed actions go into Trello with assigned owners and deadlines. Team members can update their progress and share comments.
- **Weekly Progress Check-Ins (Both)** - During standups, the team refers to a physical Kanban board in the office to track high-level progress displaying key milestones. Remote team members get updates via Slack and Notion.
- **Detailed Reporting (Digital)** - Google Sheets or Airtable digital dashboards track campaign performance and team progress, consolidating metrics and updates for stakeholders.

This hybrid flow helps you move fast, stay creative, and stay coordinated - without getting lost in the noise.

The Takeaway

Project management is part art, part system. The best ventures don't choose between digital and analog - they *curate* a blend that matches their workstyle, growth stage, and mindset.

Use digital tools for scale, automation, and collaboration. Use analog methods for focus, clarity, and creativity. And remember: the goal isn't to be perfectly organised - it's to deliver great outcomes, together.

In the next chapter, we'll walk through a real-world startup project lifecycle - start to finish - and show you how to apply everything you've learned. You'll see how to initiate, plan, execute, and close projects like a pro, with your new toolkit in hand.

5

The Startup Project Cycle: A Step-by-Step Guide

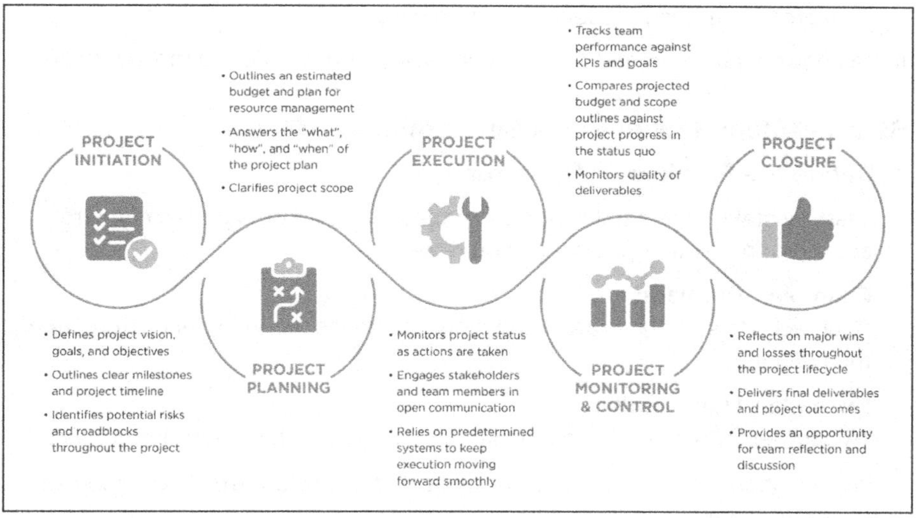

Source: Project-Management.com

---

1. Initiation: Setting the Stage for Success

Building a Strong Foundation for Your Startup Project

Every project begins with a spark - an insight, a gap in the market, or a pain point too big to ignore. But inspiration alone doesn't launch successful projects. Success starts with a solid foundation, and that foundation is built during the initiation phase. The **initiation phase** is where you lay the groundwork, ensuring your project starts with clear goals, stakeholder alignment, and a realistic plan.

In the startup world - where resources are tight and speed is essential - the initiation phase is often rushed or skipped. Don't make that mistake. This is where you define what you're doing, why it matters, who's involved, and whether the project is even viable. Get this phase right, and everything else flows more smoothly.

What Happens During the Initiation Phase?

This phase is about aligning vision and feasibility. You'll answer three core questions:

- **What are we doing?**
  Define the purpose, scope, and intended outcomes.
- **Why are we doing it?**
  Identify the business value - what will this project change, fix, or unlock?
- **Can we realistically do it?**
  Assess your team, resources, and timing.

If the answers are clear and grounded in reality, you're ready to move forward.

Step-by-Step: How to Run a Smart Initiation Phase

1. Define the Problem or Opportunity

Start by stating the catalyst for this project. It could be a problem you're solving or a market opportunity you're seizing.

Example - Problem:
"Our website's bounce rate is too high - we're losing potential customers at the first click."

Example - Opportunity:
"A competitor has exited the market, creating room for a mid-tier product."

Write a short, specific **problem or opportunity statement**. This becomes your north star.

2. Set SMART Objectives

Define what success looks like in clear, measurable terms using the **SMART framework**:

- **Specific** - Clearly define what's being done.
- **Measurable** - Set metrics for success.
- **Achievable** - Ground the objective in reality.
- **Relevant** - Align with business priorities.
- **Time-Bound** - Include a deadline.

Example:
"Redesign our homepage to reduce bounce rate by 30% within three months."

3. Conduct a Feasibility Check

Before you commit, assess whether this project is realistically doable.

Ask:

- Do we have the **budget**?
- Do we have the **people and skills**?
- Can we deliver this **within the required timeframe**?

Tip: If there are gaps, identify solutions now. That might mean reallocating internal resources or hiring a freelancer.

4. Align Key Stakeholders

Stakeholder alignment can make or break your project. Early buy-in reduces conflict, clarifies priorities, and boosts momentum.

Action Step:
Hold a project kickoff meeting. Share your objectives, scope, and expectations. Get input, build trust, and agree on success criteria.

5. Create a One-Page Project Charter

A well-crafted, concise **project charter** keeps everyone aligned without drowning them in detail.

Include:

- Project Name
- Objective Statement (SMART)
- Scope (what's included / excluded)
- Key Deliverables
- Timeline & Milestones
- Team Roles & Responsibilities
- Success Criteria

This document is your project's anchor. Revisit it regularly.

6. Validate the Idea

Before you dive into planning, validate that your idea resonates in the real world.

Validation Tactics:

- **Customer Feedback** - Interview potential users or clients.
- **Competitor Analysis** - Identify market gaps or missed opportunities.

- **Prototyping** - Build a low-fidelity prototype or landing page and test engagement.

The goal is to reduce risk and ensure you're solving the right problem the right way.

Common Pitfalls (and How to Avoid Them)

Pitfall	Solution
Rushing to Execution	Take time to clarify scope, goals, and feasibility.
Vague Objectives	Use SMART criteria to create clarity and direction.
Skipping Stakeholder Alignment	Bring everyone in early and define roles clearly.
Ignoring Resource Constraints	Be honest about bandwidth and address capacity gaps before moving forward.

Real-World Example: Meal Kit Launch

A startup launching a meal kit subscription service follows these steps:

- **Problem:** People want healthy, home-cooked meals but lack time to plan and shop.
- **Objective:** Launch the service by Q2; acquire 1,000 subscribers in the first three months.
- **Scope:** Includes product development, recipe testing, and marketing - excludes delivery (outsourced).
- **Stakeholders:** Founders, product team, marketing, logistics partner.
- **Feasibility:** Budget secured; team capacity aligned; timelines verified.
- **Validation:** Conducted customer interviews and ran a waitlist landing page.

Result: A confident go-ahead for planning and execution.

The Takeaway

The initiation phase is about focus and alignment - not doing, but deciding. By asking the right questions, defining success, and validating your idea, you build a foundation that supports every phase to come.

In the next section, we'll move into **project planning**, where we turn vision into a concrete, executable plan. Let's keep building.

2. Planning - Turning Ideas into Action

Creating a Roadmap for Success

Now that you've set the stage with a solid initiation phase, it's time to roll up your sleeves and get practical. The **planning phase** is where your project shifts from possibility to a clear, executable path. It's the moment when ideas become actions and momentum starts to build.

If initiation is about answering the big questions - "What are we doing?" and "Why are we doing it?" - then planning is all about "How are we going to make it happen?"

This section walks you through the key steps of the planning phase, startup-style - lean, focused, and designed for execution.

The Planning Process: Overview

The planning phase centers on five core activities:

1. Break It Down - define deliverables and break them into tasks
2. Set priorities
3. Build a realistic timeline with deadlines
4. Identify risks and create mitigation plans
5. Choose the right tools to stay organized

Let's break each of these down.

Step 1: Break It Down

Start with the big picture - then zoom in.

Begin by identifying your **key deliverables** - the major outputs of your project. These are the things you'll hand off, launch, or ship.

Then, break each deliverable down into clear, actionable tasks.

Example
Deliverable: Social Media Campaign
Tasks:

- Write post copy
- Design graphics
- Schedule content
- Monitor engagement

Lastly, assign ownership. Every task needs a name next to it. Avoid the trap of "we'll all chip in" - that's how things fall through the cracks.

Step 2: Set Priorities

Not everything matters equally - act accordingly.

Use tools like:

- **The Eisenhower Matrix**: Separate tasks into urgent/important categories.
- **MVP Thinking**: What's the smallest version of this project that delivers real value? Build that first.

Prioritization is how startups survive. Focus your team's energy where it matters most.

Step 3: Build a Realistic Timeline

Structure creates momentum - deadlines create accountability.

- **Identify milestones**: Key checkpoints (e.g., prototype ready, launch complete).
- **Estimate durations**: Be realistic. Add buffers.
- **Sequence the work**: What has to happen before something else can start?

Use tools like Trello, Asana, or a Google Sheets Gantt chart to visualize the timeline and track progress.

Excel Gantt Charts: Simple but Powerful Planning Tools

When building your timeline, don't overlook the power of Excel for creating visual project roadmaps. A simple Excel Gantt chart can provide clarity without the learning curve or expense of specialized software.

To create an effective Excel project timeline:

- List tasks in rows and time periods in columns
- Use cell shading to indicate task duration
- Add conditional formatting to highlight delays or dependencies
- Include milestone markers for key delivery points
- Add responsible team members for clarity

This approach combines visual clarity with the flexibility to update as your project evolves. For startups managing multiple workstreams, separate tabs can track different project components while maintaining a master timeline view.

Step 4: Identify Risks

Every startup project has unknowns - expect them.

Create a lightweight **risk register**. For each risk:

- **Describe the risk**
- **Estimate likelihood and impact**
- **Define your mitigation plan**

Example:
Risk: Key team member unavailable
Mitigation: Cross-train a second team member; document key processes

Plan for the bumps now so they don't derail you later.

Step 5: Choose Your Tools

Support your workflow - don't complicate it.

Choose tools based on your team's habits and the complexity of the project.

Recommendations:

- Trello: Visual task tracking
- Google Calendar: Shared timelines and reminders
- Slack: Communication and updates
- Notion: All-in-one docs, tracking, and wikis

Avoid overloading the team with too many platforms - start simple, and scale as needed.

Pro Tip: Apply the 80/20 Rule

Focus on what matters most.

80% of your results will likely come from 20% of your tasks. Focus your planning energy on what will drive the most value - core deliverables, key milestones, and major risks. Don't over-engineer the rest.

Example: Planning a Startup Product Launch

You're launching a new app. Here's how a lean, effective planning phase might look:

Deliverables:

- Functional MVP
- Marketing campaign
- Launch webinar

Tasks:

- Design user interface
- Build key features
- Set up landing page
- Create email sequence
- Schedule social media posts
- Book webinar platform

Milestones:

- Week 2: App prototype complete
- Week 4: Marketing ready
- Week 6: Product live

Risks & Mitigations:

- **Risk:** Developer availability
- **Mitigation:** Weekly check-ins, defined handovers

The plan is simple, clear, and adaptable - just what a startup team needs.

Pitfalls to Watch For

Pitfall	Solution
Overloading the Plan	Trying to do too much? Pull back. Stick to what's essential.
Ignoring Team Input	Your team knows the terrain. Get their input early.
Underestimating Time	Always build in buffers. Things will take longer than you think.
Skipping Risk Planning	Even a simple risk checklist can save a project.

The Takeaway

The planning phase is where vision meets execution. It's the step that transforms energy into alignment and gives your team the clarity to move forward confidently. With deliverables mapped, tasks assigned, risks anticipated, and timelines set, you're ready to execute.

In the next section, we'll enter the **execution phase** - where your plan comes to life and real progress happens. Let's keep moving..

3. Execution: Bringing Your Plan to Life

Turning Ideas into Actionable Results

Planning is where you map the path. Execution is where you walk it. This phase is where your team rolls up their sleeves and starts turning ideas into tangible results - one deliverable, one task, and one milestone at a time.

For emerging businesses, execution is often the most intense and dynamic part of the project cycle. It's when the rubber hits the road - and where things can start to go off track if you're not organized, adaptable, and communicative. But with a solid plan and a few smart habits, you can stay focused, energized, and on course.

The Goals of Execution

Execution isn't just about checking off tasks. Your goals during this phase are to:

- **Turn plans into action**: Complete tasks and deliver results as outlined.
- **Maintain momentum**: Keep the team focused, productive, and motivated.
- **Adapt to reality**: Handle surprises and course-correct without losing sight of your goals.

Step 1: Start Strong with a Kickoff

Begin execution with clarity and energy. A short kickoff meeting sets the tone for focused, aligned action.

In your kickoff, cover:

- Key goals and deliverables
- Roles and responsibilities
- Timelines and priorities
- Communication expectations
- Where to go if roadblocks arise

Pro Tip: End the meeting with a unifying rally point - like a shared milestone or mini-goal.

Step 2: Track Progress Religiously

Don't assume things are moving - make sure they are. Tracking keeps your team accountable and your project visible.

Tools and techniques:

- **Kanban boards** (Trello, Asana): Visualize what's in progress, completed, or stuck.
- **Standups or weekly check-ins**: Quick, regular updates keep communication flowing.
- **Progress metrics**: Completed tasks, milestone targets, or performance indicators.

Celebrate wins along the way. Small milestones = big morale boosts.

Step 3: Communicate Like a Pro

Projects stall when people go quiet. Keep communication active and clear, and tailor it to the audience.

Do this:

- Proactively check in before issues escalate.
- Be transparent about risks, changes, and roadblocks.
- Match the message to the receiver:
  - Team = day-to-day updates
  - Clients or investors = high-level summaries

Step 4: Stay Flexible, Stay Focused

Even the best plans hit turbulence. Execution is about reacting without overreacting.

How to stay adaptable:

- Revisit your plan weekly - adjust if needed.
- Flag blockers early and solve them fast.
- Prevent scope creep by sticking to the original plan unless a change is strategic (and agreed upon).

Step 5: Keep the Team Energized

Execution can feel like a marathon. Keep morale high by recognizing progress and supporting your team.

Ways to support and motivate:

- Acknowledge effort - publicly and often.
- Celebrate mini-milestones.
- Check in 1:1 with teammates.

- Model the mindset: optimistic, practical, and forward-looking.

Step 6: Don’t Compromise on Quality

In the push to “just get it done,” quality can slip. Build in time for review, testing, and polish.

Best practices:

- Review deliverables regularly, not just at the end.
- Test early - whether it's a new design, app feature, or marketing message.
- Capture issues and document learnings.

A Real-World Example: Executing a Marketing Campaign

Imagine your startup is launching a new product with a digital ad campaign:

- **Kickoff**: Set goals (e.g., 1,000 clicks), assign content roles, confirm tools and timeline.
- **Task tracking**: Use Trello with columns like “To Create,” “In Review,” “Scheduled,” and “Live.”
- **Check-ins**: Weekly calls to assess performance and pivot where needed.
- **Adaptation**: If ad copy underperforms, test alternatives.
- **Momentum**: Celebrate wins - “We hit 500 clicks in Week 1!”

Common Execution Pitfalls - and How to Avoid Them

Pitfall	Solution
Micromanaging	Trust your team. Focus on support, not control.
Losing sight of big goals	Re-anchor to the project objectives regularly.
Ignoring feedback	Stay open. Course-correct when new insights arise.
Team burnout	Monitor workloads and celebrate progress.

The Prevention Mindset: Spotting Issues Before They Grow

The most effective project execution isn't about fighting fires - it includes preventing them.

Throughout the execution phase, cultivate what seasoned project managers call the "prevention mindset" - actively scanning for weak signals and early warnings of potential issues. Dedicate time each week to look beyond immediate tasks and ask: "What could go wrong next?" Train yourself to notice subtle signs of trouble: slightly missed deadlines, team members who seem unusually quiet in meetings, dependencies that are approaching risk zones, or stakeholders who haven't provided expected feedback. When you spot these signals, don't wait for them to escalate into problems. Instead, have a direct conversation, adjust resources, or develop a contingency plan.

This proactive problem-solving approach transforms you from someone who merely reports issues into someone who prevents them - adding tremendous value to your team and organization.

The Takeaway

Execution is where the magic (and mess) happens. It's where good planning meets reality - and either thrives or derails. By tracking progress, communicating openly, solving problems quickly, and keeping your team energized, you'll keep momentum high and deliver results that matter.

In the next section, we'll explore the **monitoring and adjusting phase** - where you step back to fine-tune, measure, and adapt as your project moves forward. Let's keep the momentum going!

4. Monitoring and Adjusting: Staying on Track

Fine-Tuning Your Progress for Startup Success

Execution is where the rubber meets the road - but **monitoring and adjusting** is where you make sure you're still heading in the right direction. This phase is about staying aware, responsive, and nimble. Think of it as the GPS for your project - constantly checking your location and rerouting when necessary.

In a startup environment, where change is constant and speed matters, staying on track isn't about rigidly following a plan. It's about tracking progress, spotting deviations early, and making thoughtful adjustments along the way.

Why Monitoring and Adjusting Matter

This phase helps you:

- **Track Progress:** Are tasks getting done on time? Are milestones being reached?
- **Spot Problems Early:** Catch small issues before they become big blockers.
- **Stay Agile:** Respond quickly to changes in priorities, resources, or strategy.
- **Measure Impact:** Know if your project is working - and how well.

Step 1: Use Clear Tools and Metrics

You need visibility to manage effectively.

Key Metrics to Track:

- % of tasks completed vs. planned
- Milestones met vs. timeline
- Budget: actual vs. forecast
- Quality: Are deliverables meeting standards?

Tools That Help:

- Trello or Asana: Task and timeline tracking
- Monday.com: Custom dashboards and project visibility
- Google Sheets: Simple tracking for small teams
- Google Analytics, Mixpanel: For digital product or campaign performance

Excel Dashboards for Real-Time Progress Tracking

Excel's data visualization capabilities make it an excellent tool for monitoring project health with minimal setup. A well-designed Excel dashboard can provide immediate insights into your project's status.

Key elements to include:

- **Progress thermometers** showing percentage completion of key deliverables
- **Burndown charts** tracking remaining work against timeline
- **RAG (Red-Amber-Green) status** indicators for different project areas
- **Milestone tracking** with visual indicators for on-track vs. delayed items
- **Resource utilization** graphs to monitor team capacity

The advantage of Excel dashboards is their adaptability—you can easily modify them as you discover which metrics truly matter to your project's success. For remote teams, a shared Excel dashboard (via OneDrive or SharePoint) provides a single source of truth that anyone can access and update.

Pro Tip: Focus on a few critical metrics. Simplicity drives action.

Step 2: Run Regular Check-Ins

Consistency is key.

Meeting Cadence:

- Daily standups for fast-moving teams
- Weekly check-ins for larger or longer-term projects

Keep it Simple:

1. What's been completed?
2. What's in progress?
3. Are there any blockers?

Capture Next Steps: Use shared docs or comments in your PM tool to track decisions.

Step 3: Spot Deviations and Fix Them Early

Even great plans drift. The earlier you catch it, the easier it is to recover.

What to Watch For:

- Timeline slips
- Budget overages

- Missed milestones
- Task logjams or team burnout

Adjustments You Can Make:

- Shift resources (e.g., assign more help to critical tasks)
- Reorder priorities
- Re-evaluate the scope

Reminder: Not every hiccup is a crisis. Take a measured approach to course correction.

Step 4: Stay Agile and Open to Change

Startups evolve - fast. Your project needs to flex with the business.

How to Adapt Well:

- **Revisit your objectives:** Are they still relevant?
- **Re-align stakeholders:** Explain what's changing and why.
- **Update the plan:** Shift timelines, remove non-essentials, or adjust deliverables.
- **Stay focused:** Don't let reactive changes derail long-term value.

Step 5: Share Progress Transparently

Keep your team and stakeholders informed. Transparency builds trust.

Tailor Your Updates:

- **Internal team:** Daily or weekly task-focused updates
- **Stakeholders or investors:** Monthly progress summaries and key outcomes

What to Share:

- What's working
- What's off-track
- What actions you're taking to course-correct
- Wins, milestones, and learnings

Optional Visuals: A shared dashboard, timeline chart, or even a simple email summary can do the trick.

Step 6: Know When to Pivot

Sometimes, the smartest move is a strategic redirection.

Signs It's Time to Pivot:

- You're not hitting key objectives after several iterations
- A major external shift (market, funding, competitor move)
- A new opportunity has more potential than your current path

Pivot Checklist:

- Reassess goals and resources
- Communicate clearly with the team
- Update your project charter and timeline
- Preserve what you can - don't burn everything down to start over

Real-World Example: Monitoring a Website Redesign

Let's say your team is revamping your startup's website:

- **Metrics:** Monitor task completion, bounce rates, time on page
- **Tools:** Trello board for tracking, Google Analytics for site performance
- **Check-Ins:** Weekly team updates on design and content progress
- **Adjustments:** After testing, the team reallocates time to improve mobile UX
- **Communication:** Leadership receives a bi-weekly email update on progress and learnings

Common Pitfalls - and How to Avoid Them

Pitfall	Fix
Micromanaging	Trust your team. Focus on outcomes, not how people get there.
Overreacting to every issue	Stay calm. Not every deviation needs a change in direction.
Ignoring metrics	If you're not measuring, you're guessing. Track a few key things well.
Avoiding tough conversations	If something's not working - name it, fix it, move forward.

The Takeaway

Monitoring and adjusting isn't about perfection - it's about staying informed, being flexible, and keeping your eyes on the goal. By tracking progress, addressing issues early, and adapting to change, you give your startup project the best shot at success - no matter what comes your way.

Next up: **Closure** - how to wrap up a project, document your learnings, and celebrate what you've accomplished.

5. Closure: Wrapping Up and Reflecting

Celebrating Success and Capturing Lessons Learned

The finish line is in sight. You've initiated, planned, executed, and adapted your way through a startup project - and now it's time to close the loop. But closure isn't just about ticking the final checkbox. It's a strategic opportunity to deliver results, celebrate wins, and reflect on lessons that will make your next project even better.

Done well, closure strengthens your team, your systems, and your startup's learning culture. Done poorly - or skipped altogether - it can leave valuable insights (and morale) on the table.

Why the Closure Phase Matters

Startups move fast, and it's tempting to jump into the next big thing. But pausing to close your project properly pays off in several key ways:

- **Celebrate Success:** Recognition fuels motivation and retention.
- **Learn What Worked:** Reflection helps you double down on what's effective.
- **Document Lessons:** You avoid repeating mistakes by capturing insights.
- **Ensure Nothing's Missed:** Closure ensures deliverables are complete and expectations are met.

Step-by-Step Closure Checklist

1. Deliver Final Outputs

- Review final deliverables against your project objectives.
- Conduct final quality checks.
- Share deliverables with stakeholders or clients and secure their sign-off.
- Provide any supporting documentation needed for handoff.

2. Celebrate the Wins

- Publicly acknowledge the team's effort (Slack, meeting, or email).
- Host a small celebration - virtual or in-person.
- Share success stories or key metrics from the project.
- Highlight individual and team achievements.

3. Conduct a Post-Project Review

- Revisit goals: Did you meet them?

- Analyze performance across budget, timeline, and quality.
- Gather feedback from all stakeholders.
- Document wins, challenges, and insights.

Template Sections for Review:

- Project Overview
- What Went Well
- What Could Be Improved
- Lessons Learned
- Recommended Changes for Next Time

4. Archive Your Work

- Save all files and documentation to a shared, accessible space (e.g., Google Drive, Notion).
- Include: project plan, task board, deliverables, feedback, and review notes.
- Label and organize files for future reference.

5. Hold a Retrospective

- Ask your team:
    - What went well?
    - What didn't go so well?
- What should we do differently next time?
    - Keep it constructive - focus on systems, not blame.
    - Document ideas and decisions to support continuous improvement.

6. Release Resources

- Reassign or release team members to new responsibilities.
- Close out any third-party contracts or subscriptions.
- Wrap up unused budget allocations or tool access.

A Real-World Example: Wrapping Up a Product Launch

Your startup just launched a new mobile app feature. During closure, you:

- Deliver the finished app and documentation to the product team.
- Celebrate with a short Friday demo and shout-outs on Slack.
- Run a review session to analyze launch analytics.
- Archive designs, test cases, and launch assets in your shared drive.

- Hold a retrospective where the team agrees to shorten sprint cycles next time.

Common Pitfalls in the Closure Phase

Pitfall	Solution
Rushing to the next project	Schedule a firm closure window
No post-project review	Use a lightweight review template
Lack of recognition	Make celebration part of your project checklist
Lessons never captured	Document them and revisit during planning next time

The Takeaway

Project closure isn't just an afterthought - it's the final, crucial phase that transforms a completed project into a stepping stone for future success. When you deliver, celebrate, reflect, and document with intention, your team gets smarter, your process gets tighter, and your startup builds the kind of operational muscle that scales.

In the next chapter, we'll take a step back and look at how to manage **multiple projects** at once - without losing your mind or momentum. Let's finish strong and move forward even stronger.

6

Managing Small Teams and Limited Resources

---

1. The Art of Delegation

How to Empower Your Team and Maximize Efficiency

In a startup, where every team member wears multiple hats and resources are stretched thin, effective delegation isn't just a useful skill - it's essential. Delegation is the key to making the most of your limited resources while empowering your team to contribute their best work.

Yet, for many leaders, delegation can feel like a double-edged sword. On one hand, you know you can't do it all yourself. On the other, letting go of control can be challenging, especially when so much is at stake. Mastering the art of delegation means striking the right balance: assigning tasks in a way that builds trust, enhances productivity, and ensures your team thrives.

Why Delegation Matters

Emerging businesses often operate with fewer people, smaller budgets, and tighter timelines than larger organizations. Delegation allows you to:

- **Leverage Strengths:** Tap into the unique skills and expertise of each team member.
- **Save Time:** Free yourself to focus on high-priority tasks that only you can handle.
- **Build Ownership:** Empower team members by giving them responsibility and accountability.
- **Enhance Team Morale:** Show trust in your team's abilities, fostering confidence and motivation.

The Principles of Effective Delegation

1. Know What to Delegate

Not every task is a good candidate for delegation. Focus on tasks that:

- Don't require your unique expertise or decision-making.
- Can help team members grow their skills.
- Are repetitive or time-consuming for you but important to the project.

Examples:

- Delegate: Data analysis, social media posting, event logistics.
- Don't Delegate: Critical strategy decisions, high-stakes negotiations, performance reviews.

2. Match Tasks to the Right People

Delegation works best when tasks align with a team member's skills, experience, and interests.

Steps to Match Tasks:

- Assess your team's strengths and areas for growth.
- Assign tasks that leverage existing skills or offer opportunities for development.
- Consider workload balance to avoid overwhelming any single person.

3. Communicate Clearly

When delegating a task, clarity is crucial. Your team can't deliver what you want if they don't understand the expectations.

What to Communicate:

- **The Task:** What needs to be done, and why it matters.
- **The Deadline:** When it needs to be completed.
- **The Standards:** What success looks like (quality, format, outcomes).
- **The Resources:** What tools, support, or information they'll need.
- **The Autonomy Level:** How much decision-making freedom they have.

4. Empower Your Team

Delegation isn't about micromanaging - it's about trust. Once you've assigned a task, give your team the space to execute it in their way.

Ways to Empower:

- Encourage ownership and responsibility.
- Offer guidance and support, but avoid hovering.
- Be available for questions, but resist taking back control.

5. Monitor Progress Without Micromanaging

Stay informed without undermining your team's autonomy.

How to Monitor Effectively:

- Set regular check-ins or updates.
- Use tools like Trello or Asana to track progress visually.
- Focus on results, not every step of the process.

6. Provide Feedback and Recognition

Delegation is also a chance to coach your team. Feedback helps them improve, while recognition builds morale.

Tips for Feedback:

- Be specific - call out what went well and what can improve.
- Tie feedback to impact.
- Celebrate small and large wins - both matter.

Overcoming Delegation Challenges

Reluctance to Let Go

- **Problem:** Fear that the task won't be done "right."
- **Solution:** Focus on the outcome, not the process. Trust your team.

Lack of Time to Delegate

- **Problem:** It feels quicker to do it yourself.
- **Solution:** Invest upfront. Delegation pays off over time.

Uneven Work Distribution

- **Problem:** Some team members feel overloaded.
- **Solution:** Regularly assess and rebalance workloads.

Unclear Expectations

- **Problem:** Tasks fall short due to miscommunication.
- **Solution:** Be clear, confirm understanding, and document expectations.

Real-World Example: Delegation in Action

A startup founder is juggling product development, investor relations, and a marketing push. Here's how they delegate:

- **Task:** Organize a product launch event
  Assigned to: Operations manager
- **Task:** Write social media copy and press materials
  Assigned to: Marketing intern
- **Task:** Coordinate with beta testers
  Assigned to: Product manager

By providing context, empowering the team, and following up at key milestones, the founder is able to focus on strategic priorities while the team drives execution forward.

Creating a Team of Horizon Scanners

The best project leaders don't just organize their teams - they empower them to become horizon scanners too. Create a culture where team members feel responsible for identifying potential issues in their areas of expertise and are rewarded - not punished - for surfacing concerns early.

During standups and check-ins, don't just ask about completed tasks; specifically prompt for potential roadblocks team members see coming. Simple questions like "What might prevent you from meeting next week's milestone?" or "What dependencies are you concerned about?" train your team to think proactively.

When you transform your entire team into problem anticipators, you multiply your effectiveness exponentially. In startups where everyone wears multiple hats, this distributed vigilance becomes your secret weapon for keeping projects on track despite limited resources.

The Takeaway

Delegation isn't about losing control - it's about building trust and scaling your impact. When done well, it develops your team's capabilities, prevents burnout, and allows you to stay focused on what matters most. In your startup, learning to delegate with intention may be the most powerful productivity tool you have.

2. Maximizing Limited Budgets

Making Every Dollar Count in Your Startup

Startups are the ultimate exercise in doing more with less. Limited budgets are the norm, and every dollar you spend needs to stretch further than a yoga instructor on retreat. But here's the good news: constraints can spark creativity. With the right strategies, a lean budget doesn't have to mean limited impact.

In this section, we'll explore how to make the most of every penny, prioritize spending on what truly matters, and find creative ways to achieve big results on a small budget.

Why Budget Optimization Matters

A tight budget isn't just a challenge - it's a competitive advantage. It forces you to:

- **Focus on Efficiency:** Cut unnecessary costs and streamline operations.
- **Prioritize with Purpose:** Spend only where you'll get real returns.
- **Drive Innovation:** Limited funds push you toward resourceful, creative solutions.
- **Build Financial Discipline Early:** Setting smart habits now pays off when your budget grows later.

Step 1: Start with a Budget Blueprint

A clear, simple budget helps you see where every dollar goes.

How to Set It Up:

- **Define Categories:** Group expenses (e.g., marketing, tools, salaries, operations).
- **Assign Spending Limits:** Allocate a budget for each based on project priorities.
- **Track Every Dollar:** Use Google Sheets, Airtable, or QuickBooks to monitor in real time.

Example - Startup Marketing Budget:

- Content Creation: $2,000
- Ad Spend: $1,500
- Tools/Design Software: $500
- Contingency Fund: $300

Pro Tip: Always include a contingency (10-15%) for unexpected costs.

Excel: The Budget-Friendly Financial Management Tool

When managing your project finances, Excel offers sophisticated budget tracking capabilities without additional software costs. For startups watching every dollar, an Excel budget template can be customized to your exact needs.

Essential Excel budget features to leverage:

- **Expense forecasting** with formula-based projections
- **Cash flow visualization** to prevent timing problems
- **Scenario planning** using data tables to model different outcomes
- **Budget vs. actual comparisons** with automatic variance calculations
- **Conditional formatting** to highlight areas needing attention

Pro tip: Create a master budget template that links to individual project budgets, giving you both detailed control and big-picture insights. This approach scales with your startup, eliminating the need for expensive financial software until your operations truly require it.

Step 2: Prioritize High-Impact Spending

Every startup has "must-haves" and "nice-to-haves." Focus your budget where it counts.

Ask Yourself:

- Does this directly support our business goal?
- Can we get the same result more affordably?
- What's the ROI?

Framework:

- **Essential:** Tools, software, services critical to your operations.
- **High Impact:** Things that significantly improve outcomes.
- **Nice-to-Have:** Save for later or only fund if the budget allows.

Step 3: Embrace DIY Solutions

You don't need to outsource everything. Many tasks can be handled in-house with free or low-cost tools.

DIY Ideas:

- **Design:** Canva, Figma, or Adobe Express

- **Marketing:** Mailchimp (free tier), Buffer, LinkedIn posts
- **Surveys/Research:** Google Forms, Typeform (free version)

Know when not to DIY: Legal advice, taxes, or product safety should usually be handled by professionals.

Step 4: Leverage Free and Affordable Tools

There's an ecosystem of budget-friendly tools made just for small teams.

Free Tools to Explore:

- **Project Management:** Trello, Asana, Notion
- **Collaboration:** Google Workspace, Slack, Zoom
- **Marketing/Design:** Canva, Mailchimp, Buffer
- **Analytics:** Google Analytics, Mixpanel

Many offer affordable startup plans with more features if you need to scale.

Step 5: Barter, Trade, or Partner

When cash is tight, creativity matters.

Ways to Trade Value:

- **Barter Services:** Swap skills with other startups. (You do branding, they help with SEO.)
- **Collaborate:** Co-market with other businesses to share costs and increase reach.
- **Use Your Network:** Ask mentors, advisors, or alumni for resources, advice, or referrals.

Step 6: Negotiate Everything

Startup founders often overlook the power of negotiation.

Negotiation Tips:

- Ask about startup discounts or nonprofit rates.
- Offer testimonials or referrals in exchange for better pricing.
- Bundle services to save (e.g., annual subscriptions vs. monthly).
- Pitch your growth story - vendors love future potential.

Step 7: Track and Review Weekly

A budget only works if you actively manage it.

Budget Management Habits:

- **Weekly Reviews:** Check expenses every week - don't wait until the end of the month.
- **Watch for Red Flags:** Overages, duplicate subscriptions, or unused tools.
- **Refine as You Go:** Compare actual vs. projected spend and reallocate accordingly.

Real-World Example: Hosting an Event on a Shoestring Budget

Goal: Run a networking event with only $2,000.

How They Did It:

- **Venue:** Used a free co-working space in exchange for promotion.
- **Food/Drinks:** Provided self-serve snacks from Costco.
- **Promotion:** Designed invites in Canva and promoted via LinkedIn and email lists.
- **Photography:** Traded event access for photos with a local creative.

Outcome: Professional-looking event, great feedback, new business leads - and stayed under budget.

Common Pitfalls and How to Avoid Them

Pitfall	Solution
Overspending on "Nice-to-Haves"	Stick to core priorities tied to measurable outcomes
Ignoring the ROI	Regularly evaluate what you're getting for your spend
No Contingency Plan	Always set aside funds for the unexpected
Forgetting About "Small" Costs	Subscription and transaction fees add up - track everything

The Takeaway

Lean budgets are your reality - but they don't have to be a liability. With smart prioritization, creative thinking, and disciplined management, you can do more with less and set your startup up for long-term financial strength. Constraints can lead to innovation - and your limited budget might just be your secret weapon.

3. Managing Remote and Hybrid Teams

Building Collaboration Across Locations

Remote and hybrid teams are no longer just a trend - they've become standard in today's startup and emerging business environments. Whether your team is fully distributed or working across time zones a few days a week, leading a project in this environment brings new challenges - and opportunities.

Project managers must adapt their leadership style to maintain alignment, encourage collaboration, and support team members who may never share a physical workspace. The good news? When managed well, remote teams can be just as productive - if not more so - than traditional in-office teams.

Clarify Expectations Early

The foundation of successful remote work is clarity. Without the casual check-ins of an office environment, ambiguity can derail progress quickly.

- **Set clear goals and deadlines** for every phase of the project.
- **Define roles and responsibilities** so team members know what they own.
- **Clarify communication channels** - who needs to know what, when, and how often.

Pro Tip: Use a simple project overview document that outlines expectations, communication norms, and meeting cadences. Make this a living document your team can reference.

Create a Shared Workflow

Remote teams benefit from having clearly structured collaboration workflows. Whether through daily standups, weekly syncs, or shared boards in Notion or Trello, creating a rhythm helps everyone understand where the project stands and who's doing what. Defining steps and substeps for common processes allows tasks to flow smoothly between team members and avoids the risk of bottlenecks.

Use the Right Tools - And Use Them Well

Remote teams rely on tools to stay connected and on track. Your toolkit might include:

- **Communication**: Slack, Microsoft Teams, Zoom
- **Project Management**: Trello, Asana, Monday.com, Notion
- **Documentation**: Google Docs, Confluence, Dropbox Paper

- **Scheduling**: Calendly, Doodle, World Time Buddy

Be Strategic About Communication Channels: Not all communication tools serve the same purpose. Instant messaging is ideal for quick updates, video calls are best for in-depth discussions, and email serves more formal or documented exchanges. Establishing norms around when to use each channel can prevent miscommunication and streamline collaboration.

Pro Tip: Use asynchronous tools when possible to respect time zone differences and reduce meeting fatigue.

Hold Regular Check-Ins

Consistency builds trust. Whether it's a quick daily huddle or a more in-depth weekly review, check-ins help keep everyone aligned and provide space for questions or challenges to surface.

- Start with a personal check-in: "How are you doing?"
- Then move to project updates: "What's going well? What's getting in the way?"
- End with support: "What do you need from me or the team?"

Pro Tip: Encourage camera-on meetings at least once a week to maintain a human connection.

Foster a Culture of Trust and Accountability

Remote teams thrive when there's a balance between autonomy and accountability. Micro-managing doesn't work across screens - but neither does radio silence.

- Empower team members to manage their time, but ensure expectations are well defined.
- Use shared task boards or trackers so progress is visible to everyone.
- Celebrate follow-through and hold people accountable to commitments.

Offer Flexibility with Accountability: Studies show that employees with flexible schedules are significantly more focused and productive than those on rigid ones. Instead of fixed hours, consider allowing team members to choose when they work - provided they meet expectations and remain available for key touchpoints.

Encourage Team Connection

Just because you're not in the same office doesn't mean you can't build rapport.

Create Opportunities for Casual Connection: Building genuine human connection is essential for remote team cohesion. Consider organizing casual, low-stakes virtual events - such as monthly game nights or themed coffee chats. These informal moments help employees build trust, share experiences, and strengthen their ability to collaborate effectively when it counts.

Pro Tip: Use Slack channels or Teams groups for informal chatter, hobbies, or shout-outs.

Adapt to Different Communication Styles

Just as people have different working styles, they also vary in how they receive and process information. Some team members prefer detailed instructions; others thrive with a high-level overview and room to explore. As a leader, being able to flex your style depending on who you're working with enhances clarity and makes your communication more effective across the board.

Remote Teams Need Proactive Leadership

Managing a remote or hybrid team isn't about "setting and forgetting." It takes deliberate effort to build connection, maintain clarity, and support productivity. With intentional practices like structured workflows, flexible scheduling, and tailored communication, you can unlock the full potential of your remote team.

Common Challenges and How to Overcome Them

1. **Isolation and Burnout:**
   - **The Problem:** Remote workers may feel disconnected and overworked.
   - **The Solution:** Encourage regular breaks, foster social interaction, and promote a culture of openness and support.
2. **Communication Gaps:**
   - **The Problem:** Important information gets lost or misunderstood.
   - **The Solution:** Use consistent communication channels and ensure everyone has access to updates.
3. **Uneven Engagement:**
   - **The Problem:** Hybrid team members in the office may feel more connected than remote workers.
   - **The Solution:** Include remote team members in all decisions, discussions, and celebrations to ensure equity.
4. **Technology Fatigue:**
   - **The Problem:** Too many tools or constant notifications overwhelm the team.
   - **The Solution:** Simplify your tech stack and encourage "offline" work time.

Real-World Example: Managing a Hybrid Product Development Team

Imagine a startup with a hybrid team developing a new app. Here's how they manage the process:

1. **Communication:**
   - Use Slack for quick updates and Notion for centralized project documentation.
   - Hold weekly team meetings on Zoom to discuss progress and resolve issues.
2. **Task Management:**
   - Use Asana to assign tasks, track deadlines, and share progress updates.
3. **Team Connection:**
   - Schedule biweekly virtual happy hours to maintain a sense of camaraderie.
4. **Time Zones:**
   - Developers in different regions work asynchronously, sharing updates via recorded Loom videos.

By combining clear communication, effective tools, and a culture of trust, the team successfully launches their app on schedule while maintaining high morale.

The Takeaway

Managing remote and hybrid teams requires flexibility, strong communication, and a focus on connection. By leveraging the right tools, fostering trust, and addressing challenges proactively, you can build a cohesive, productive team - regardless of where they're working.

In the next chapter, we'll explore advanced strategies for managing uncertainty and navigating pivots, ensuring your startup thrives in an ever-changing environment. Let's keep building together!

7

Mastering Time Management in Startups

---

1. Prioritizing What Matters Most

Focus on What Drives Results

In a startup, time is your most precious resource. With endless tasks, tight deadlines, and limited manpower, it's easy to feel overwhelmed. But here's the secret: you don't have to do everything. Successful time management isn't about cramming more into your day - it's about focusing on what truly matters.

Prioritization is the art of separating the urgent from the important, the nice-to-haves from the must-haves. By mastering this skill, you can maximize your impact, reduce stress, and keep your startup moving toward its goals.

Why Prioritization Matters in Startups

Startups thrive in an environment of uncertainty and rapid change. Without clear prioritization, teams can waste time on low-value activities while critical tasks fall through the cracks. Prioritization helps you:

1. **Stay Focused:**
    - Concentrate on tasks that directly contribute to your goals.
2. **Use Resources Wisely:**
    - Allocate time, energy, and attention where they'll have the most impact.
3. **Maintain Momentum:**
    - Keep your team aligned and moving forward, even when priorities shift.
4. **Reduce Stress:**
    - Eliminate the mental clutter of trying to do it all.

The Frameworks for Prioritizing Work

Several tried-and-true frameworks can help you prioritize effectively. Here's how to apply them in a startup context:

1. The Eisenhower Matrix

This classic tool divides tasks into four quadrants based on urgency and importance:

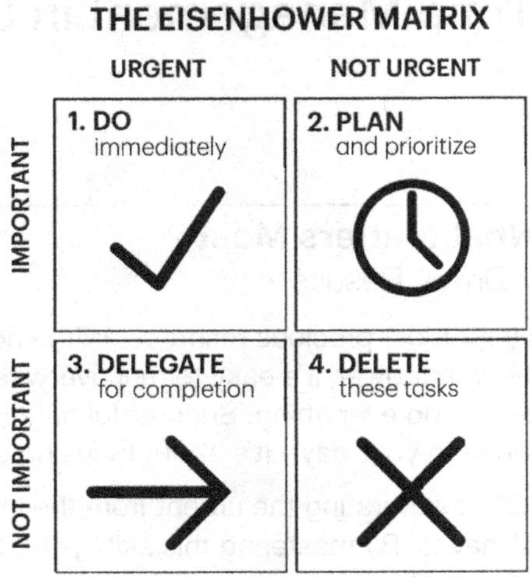

- **Quadrant 1:** Urgent & Important - Do these now (e.g., resolving critical bugs, meeting deadlines).
- **Quadrant 2:** Not Urgent & Important - Schedule these (e.g., strategic planning, team development).
- **Quadrant 3:** Urgent & Not Important - Delegate these (e.g., routine admin tasks).
- **Quadrant 4:** Not Urgent & Not Important - Eliminate these (e.g., low-priority emails, unnecessary meetings).

How to Use It:
At the start of each week or day, categorize your tasks into these quadrants. Focus on Quadrants 1 and 2 to balance immediate needs with long-term goals.

2. The 80/20 Rule (Pareto Principle)

The Pareto Principle states that 80% of results come from 20% of efforts. Identify the tasks, projects, or activities that deliver the most value and prioritize them.

Example:
If customer retention drives 80% of your revenue, prioritize tasks like improving user experience or crafting loyalty programs over chasing new leads.

3. The One Thing

This approach, inspired by the book *The One Thing* by Gary Keller, encourages you to focus on the single most impactful task at any given time.

Ask Yourself:

- "What's the one thing I can do today that will make everything else easier or unnecessary?"

Example:
If your startup's goal is to launch a product, your "one thing" might be finalizing the MVP.

Steps to Prioritize Effectively

1. Clarify Your Goals

Start by identifying your startup's top priorities. What are the key objectives for this week, month, or quarter?

Pro Tip: Use SMART goals (Specific, Measurable, Achievable, Relevant, Time-bound) to create clarity.

2. Break Down Big Tasks

Large goals can feel overwhelming. Break them into smaller, actionable tasks and prioritize based on dependencies and deadlines.

Example:
If your goal is to launch a marketing campaign, break it down into tasks like creating content, setting up ads, and analyzing performance metrics.

3. Evaluate Tasks Against Impact and Effort

Use an **Impact/Effort Matrix** to assess which tasks deliver the most value for the least effort.

- **High Impact, Low Effort:** Do these first.
- **High Impact, High Effort:** Schedule and allocate resources.
- **Low Impact, Low Effort:** Delegate or do when you have downtime.
- **Low Impact, High Effort:** Eliminate or deprioritize.

4. **Reassess Regularly**

   Priorities can shift quickly in startups. Schedule regular check-ins (e.g., weekly or biweekly) to reassess your priorities and adjust as needed.

Common Pitfalls and How to Avoid Them

1. **Mistaking Urgency for Importance:**
   - **The Problem:** Spending time on tasks that feel urgent but don't contribute to long-term goals.
   - **The Solution:** Use the Eisenhower Matrix to differentiate between urgency and importance.
2. **Overcommitting:**
   - **The Problem:** Taking on too much, leading to burnout and missed deadlines.
   - **The Solution:** Be realistic about what you can accomplish and say "no" to low-priority tasks.
3. **Neglecting Long-Term Goals:**
   - **The Problem:** Focusing only on immediate needs while ignoring strategic planning.
   - **The Solution:** Block time for Quadrant 2 activities (important but not urgent).
4. **Lack of Clarity:**
   - **The Problem:** Team members are unclear about priorities, leading to confusion and wasted effort.
   - **The Solution:** Clearly communicate priorities and update them as needed.

Real-World Example: Prioritizing in a Startup

Imagine a startup preparing for an investor pitch while simultaneously working on a product launch. Here's how they prioritize:

1. **Clarify Goals:**
   - Secure funding (top priority).
   - Deliver a polished MVP for user testing.
2. **Break Down Tasks:**
   - For the pitch: Create a slide deck, rehearse presentations, and finalize financial projections.

- For the MVP: Complete key features, fix critical bugs, and gather user feedback.

3. **Apply the Eisenhower Matrix:**
   - Urgent & Important: Rehearsing for the pitch, fixing critical bugs.
   - Not Urgent & Important: Scheduling follow-ups with potential investors.
   - Urgent & Not Important: Responding to general inquiries (delegate).
   - Not Urgent & Not Important: Refining the website's design (deprioritize).
4. **Reassess Weekly:**
   - Adjust focus after the pitch to prioritize the MVP launch.

By staying focused on high-impact tasks, the startup successfully secures funding and launches its product on time.

The Takeaway

In startups, effective prioritization is the key to turning limited time into meaningful results. By focusing on what matters most and regularly reassessing your priorities, you can ensure your efforts align with your goals. Prioritization isn't about doing everything - it's about doing the right things.

In the next section, we'll explore strategies for minimizing distractions and staying productive in the face of startup chaos. Let's keep moving forward!

2. Managing Distractions and Staying Focused

Finding Clarity Amid Startup Chaos

Startups are buzzing with ideas, energy, and opportunities - but they're also rife with distractions. From constant notifications to last-minute changes in priorities, staying focused in this environment can feel like an uphill battle. The ability to tune out distractions and zero in on what matters most is a superpower for startup founders and teams alike.

This section explores practical strategies to minimize distractions, create focused work environments, and maintain productivity - even in the chaotic world of startups.

Why Focus Matters in Startups

In a startup, every minute counts. Distractions can derail progress, waste valuable time, and lead to burnout. Staying focused ensures you:

1. **Maximize Impact:**
    - Spend your time on activities that drive real results.
2. **Reduce Stress:**
    - Avoid the mental fatigue of constantly switching between tasks.
3. **Boost Efficiency:**
    - Accomplish more in less time by working with intention.
4. **Maintain Clarity:**
    - Keep your team aligned and moving toward shared goals.

Common Startup Distractions

Understanding what pulls your attention away is the first step to combating distractions. Here are some common culprits:

1. **Constant Notifications:**
    - Emails, Slack messages, and app alerts can interrupt deep work.
2. **Unclear Priorities:**
    - Without a clear direction, it's easy to get sidetracked by low-value tasks.
3. **Multitasking:**
    - Juggling too many tasks at once reduces overall productivity.
4. **Last-Minute Requests:**
    - Unexpected demands can derail your planned workday.

5. **Open-Office or Remote Chaos:**
    - Noise, interruptions, or a cluttered environment can make it hard to concentrate.

Strategies for Managing Distractions

1. **Set Boundaries with Technology**

    Technology is both a blessing and a curse. Learn to control it before it controls you.

    Tips to Minimize Digital Distractions:

    - **Turn Off Notifications:** Silence non-essential alerts during focused work periods.
    - **Use "Do Not Disturb" Modes:** Block notifications on devices during meetings or deep work sessions.
    - **Batch Communication:** Check emails and messages at set times instead of constantly monitoring them.
    - **Leverage Focus Apps:** Use tools like Focus@Will or Forest to create distraction-free work periods.

2. **Master Your Environment**

    A well-designed workspace helps you concentrate and stay productive.

    For Office Settings:

    - Use noise-canceling headphones to block out distractions.
    - Create "focus zones" where interruptions are discouraged.
    - Keep your desk organized and free of clutter.

    For Remote Work:

    - Designate a dedicated workspace, even if it's just a corner of a room.
    - Communicate availability to others at home to minimize interruptions.
    - Stick to a consistent schedule to create work-life balance.

3. **Embrace Time Blocking**

    Time blocking is a powerful technique for structuring your day.

    How to Use Time Blocking:

    - Break your day into chunks dedicated to specific tasks.
    - Reserve blocks for deep work, admin tasks, and breaks.
    - Honor your time blocks like appointments - don't let distractions intrude.

Pro Tip: Use tools like Google Calendar or apps like Toggl to plan and track your time blocks.

4. **Manage Multitasking Temptations**

   Multitasking can feel productive but often leads to mistakes and wasted time.

   How to Avoid Multitasking:

   - Focus on one task at a time, completing it before moving to the next.
   - Use a task list to keep track of what's next, reducing the urge to jump around.
   - Practice the "Two-Minute Rule": If a quick task arises, handle it immediately or defer it.

5. **Prioritize Regular Breaks**

   Breaks aren't distractions - they're essential for maintaining focus and preventing burnout.

   Best Practices for Breaks:

   - Follow the **Pomodoro Technique**: Work for 25 minutes, then take a 5-minute break.
   - Step away from screens during breaks - stretch, hydrate, or take a short walk.
   - Schedule longer breaks during the day to recharge and refocus.

Staying Focused in Real Time

Even with the best plans, distractions will arise. Here's how to stay focused in the moment:

1. **Pause and Evaluate:**
   - Ask, "Does this interruption require immediate attention, or can it wait?"
2. **Deflect Non-Urgent Requests:**
   - Politely but firmly redirect interruptions to a later time.
   - Example: "I'm in the middle of something right now - can we discuss this after 3 PM?"
3. **Practice Mindfulness:**
   - Stay present by focusing on one task at a time.
   - Use mindfulness techniques like deep breathing to regain focus if you feel overwhelmed.

Common Challenges and How to Overcome Them

1. **Feeling Overwhelmed by Notifications:**
   - **The Problem:** Constant pings create a sense of urgency.
   - **The Solution:** Set specific times to check emails and messages, and turn off non-essential alerts.
2. **Unpredictable Interruptions:**
   - **The Problem:** Ad-hoc requests disrupt focus.
   - **The Solution:** Create "office hours" when you're available for questions or discussions.
3. **Inconsistent Work Habits:**
   - **The Problem:** Lack of routine makes it harder to stay focused.
   - **The Solution:** Establish a daily schedule that includes focused work periods, breaks, and admin time.

Real-World Example: Managing Distractions in a Startup

Imagine a founder balancing product development with investor outreach. Here's how they stay focused:

1. **Time Blocking:**
   - Reserve mornings for uninterrupted deep work on product development.
   - Dedicate afternoons to responding to emails, holding meetings, and engaging with investors.
2. **Setting Boundaries:**
   - Turn off Slack notifications during deep work periods.
   - Use a shared calendar to signal availability for team discussions.
3. **Minimizing Distractions:**
   - Work in a quiet, dedicated workspace with a clear desk.
   - Use noise-canceling headphones and the Focus@Will app for concentration.
4. **Regular Breaks:**
   - Take a 10-minute walk between time blocks to recharge and refocus.

By structuring their day and minimizing distractions, the founder stays productive and maintains clarity amid startup demands.

The Takeaway

In a fast-paced startup environment, distractions are inevitable - but they don't have to derail your productivity. By creating a focused workspace, setting boundaries with technology, and embracing time management techniques, you can maintain clarity and drive results. Focus isn't about eliminating every interruption; it's about managing your attention with intention.

In the next section, we'll explore strategies for staying productive under pressure, ensuring you and your team thrive even during high-stress periods. Let's keep moving forward!

8

Dealing with Uncertainty and Pivots

---

1. Embracing Change

Thriving in the Unpredictable World of Startups

In the fast-paced and ever-changing world of emerging businesses, uncertainty is the norm - not the exception. Markets shift, competitors appear out of nowhere, customer needs evolve, and the best-laid plans often require course correction. For founders, embracing change isn't just a survival tactic - it's a competitive edge.

Startups that thrive in chaos do so because they *expect* change. They don't see it as a threat but as an opportunity to adapt, innovate, and lead. This section offers practical strategies to help you build agility into your mindset, team, and operations - so you're ready for whatever comes next.

Why Embracing Change Matters

Change, when embraced, becomes a catalyst for progress. In the fast-moving world of startups, your ability to adapt determines whether you lead the market - or get left behind.

Key Benefits of Embracing Change:

- **Adapt Quickly:** Stay responsive to new challenges and opportunities.
- **Innovate Faster:** Use feedback and market shifts to fuel product evolution.
- **Build Resilience:** Gain confidence in navigating future uncertainty.
- **Enhance Customer Relevance:** Align faster with evolving customer needs.

Techniques for Thriving in Change

1. Cultivate a Growth Mindset

A growth mindset helps you see challenges as learning opportunities - not failures.

How to do it:

- Encourage your team to reflect on what they're learning - not just what they're achieving.
- Reframe mistakes as insights.
- Lead by example. When change hits, demonstrate flexibility and curiosity.

Prompt: Ask your team in retrospectives: "What did we learn this week that surprised us?"

2. Stay Anchored in Vision, Not in Plans

Adaptation doesn't mean abandoning your mission - it means finding better paths to get there.

How to stay grounded:

- Regularly revisit your startup's *vision* and *why*.
- Allow flexibility in *how* you achieve it.
- Evaluate each new opportunity or challenge by asking: "Does this align with our mission?"

3. Make Iterative Adjustments

Big pivots get headlines - but small, continuous iterations drive real growth.

How to iterate:

- Test small changes based on feedback or performance dips.
- Launch early, test fast, improve often.
- Track key indicators (user behavior, engagement, churn) to guide your next move.

4. Build a Team That Adapts

Agile teams are your most powerful asset in times of change.

What to focus on:

- Hire for adaptability, not just credentials.
- Cross-train roles to build flexibility into your team.
- Encourage shared ownership of problems and solutions.

Pro Tip: Hold "idea sprints" during uncertain periods to unlock creative solutions from across your team.

5. Use Data to Stay Grounded

Data turns uncertainty into insight. Don't guess - measure, analyze, then act.

Where to look:

- Customer behavior changes (e.g., drop in engagement or usage).
- Industry trends and competitor pivots.
- Internal metrics like team velocity, campaign ROI, or retention curves.

6. Communicate Transparently

When change happens, silence breeds anxiety. Transparency builds trust.

What to communicate:

- Why the change is happening.
- How it affects the team, customers, or strategy.
- What the plan is moving forward.

How to communicate:

- Use roadmaps or infographics to make complex updates clear.
- Acknowledge what's unknown - but show confidence in the process.

Common Challenges and How to Overcome Them

Challenge	Solution
Team Resistance to Change	Involve them early; explain the "why" behind changes
Decision Paralysis	Make small, reversible decisions and iterate
Overreacting to Uncertainty	Stay grounded in your vision. Use data and feedback to inform changes

Real-World Example: Navigating a Market Shift

A startup offering on-demand meal delivery notices a decline in orders as users return to cooking at home. Rather than doubling down on ads, they explore user feedback and test a low-risk pilot: offering ready-to-cook meal kits.

- **Step 1:** Collect user feedback through in-app surveys.
- **Step 2:** Launch a 30-day meal kit pilot using existing infrastructure.
- **Step 3:** Monitor sales, retention, and cost metrics.
- **Step 4:** Reallocate the budget to grow the new line after early traction.

The result? A new revenue stream and a stronger business model aligned with changing user needs.

The Takeaway

Change is inevitable in startups - but it doesn't have to be scary. When you embrace change as a natural part of the journey, you unlock your team's creativity, adaptability, and resilience. Stay flexible, grounded in purpose, and open to learning - and your startup won't just survive uncertainty. It'll thrive in it.

Up next: How to recognize *when* it's time to pivot - and how to do it without losing momentum.

2. When to Pivot

Signs It's Time to Change Course - And How to Do It Gracefully

In the startup world, resilience is often praised. Grit gets romanticized. But here's the truth: there's a fine line between persistence and denial. Sometimes the bravest, most strategic move you can make isn't to keep pushing forward - it's to pivot.

Success is rarely a straight line. Often, the willingness to pivot - making a significant change in direction - determines whether a startup thrives or falters. Pivoting isn't about admitting failure; it's about recognizing that the current path isn't leading to your destination and having the courage to chart a new course.

A pivot isn't a sign of failure. Based on what you've learned, it's a strategic shift - usually in product, market, or business model. But how do you know when it's time to pivot? Pivoting too soon can waste time and resources, while waiting too long can lead to missed opportunities. Knowing ***when*** to pivot and ***how*** to do it with clarity and momentum is the key.

What Is a Pivot, Really?

A pivot is a deliberate shift in strategy to address a fundamental issue or seize a new opportunity. It might involve changing your product, target audience, business model, or go-to-market strategy.

Common Pivot Types:

- **Product Pivot** - Changing the features or focus of your product.
- **Market Pivot** - Targeting a different customer segment.
- **Business Model Pivot** - Shifting how you generate revenue (e.g., freemium → subscription).
- **Technology Pivot** - Using different tech or infrastructure for better scalability or efficiency.
- **Channel Pivot** - Finding new ways to reach or distribute to your customers.

Step 1: Spot the Signs It's Time to Pivot

Not every setback requires a pivot. But when the same signals keep showing up, it's time to pay attention.

Key Signs to Watch:

- **Stagnant Growth**
  Revenue, usage, or engagement flatlines despite repeated efforts to improve.

- **Customer Misalignment**
  Users don't "get it" or are asking for something you're not offering.
- **Negative Retention**
  People try your product... and never come back.
- **Burn Rate Pressure**
  You're running out of runway without traction to show for it.
- **Team Misalignment**
  If your team no longer believes in the vision, that's a red flag.
- **Emerging Opportunities**
  A new market, segment, or use case is showing unexpected traction.

Listen to your gut - but back it with data. If something feels off, dig deeper into the numbers and feedback loops.

Step 2: Evaluate Before You Pivot

Before you make a big move, take a beat. Pivots are disruptive - and potentially expensive - so make sure it's the right choice.

Ask These Questions:

- Have we tried enough iterations on our current path?
- Is there a clear problem-solution mismatch?
- Are customers signaling a better direction?
- Can we pivot without losing our core strengths?
- Do we have enough resources and buy-in to make the shift?

Pro Tip: Bring your team and advisors into this process early. A pivot done in isolation is risky - a pivot done with support builds momentum.

Step 3: Plan the Pivot

Once you decide to pivot, don't wing it. A well-executed pivot feels almost surgical: focused, deliberate, and based on insight - not panic.

Planning Steps:

1. **Define the New Direction**
   What are you shifting *to*, and why?
2. **Set Metrics for Success**
   How will you know the pivot is working (e.g., retention, revenue per user)?
3. **Build a Roadmap**
   Break down the pivot into clear phases - test, learn, scale.

4. **Communicate Clearly**
   Internally and externally, explain the "why" behind the pivot.

5. **Reallocate Resources**
   Shift team focus, budget, and energy where it counts.

Step 4: Execute Without Chaos

Pivots can energize your startup - or derail it. Keep things tight, transparent, and focused.

Execution Tips:

- **Start Small:** Pilot your new direction with a subset of users.
- **Use What You've Got:** Don't throw away what's working - bring your strengths with you.
- **Stay Close to Customers:** Keep validating your assumptions with real feedback.
- **Manage Morale:** Change is emotional. Keep your team engaged and included.

Step 5: Reflect, Learn, and Iterate

Pivots are part of a learning journey. Whether they skyrocket your growth or just get you one step closer to clarity, they're valuable.

After the Pivot:

- Analyze pre- and post-pivot metrics.
- Hold a retrospective with your team.
- Document lessons learned.
- Celebrate progress - even if you're still iterating.

Real-World Example: The Classic Slack Pivot

Slack didn't start as Slack. It was born from the ashes of a failed game company.

- **The Pivot:** From gaming to workplace communication.
- **Why It Worked:**
  - They noticed internal tools solved a real pain point.
  - Validated the need with early users.
  - Built a beta and scaled from there.
  - Clear vision, great execution, relentless iteration.

- **Result:** A billion-dollar SaaS company used by millions of teams.

Common Pitfalls to Avoid

Pitfall	Solution
Pivoting out of panic	Gather data. Pause. Plan.
Changing too many things	Focus on one major change at a time.
Failing to communicate	Bring your team, customers, and investors along for the ride.
Ignoring cultural impact	A pivot changes your story - be ready to rewrite it together.

The Takeaway

Knowing *when* to pivot - and how to do it gracefully - is a hallmark of a strong founder. When your current path isn't getting results, it's not weakness to change course. It's leadership.

Done right, a pivot isn't a detour. It's a better route to your destination.

In the next section, we'll explore how to lead your team through the emotional and operational ups and downs of uncertainty - because mindset and morale are just as important as strategy.

3. Managing Stress and Setbacks

Staying Resilient in the Face of Challenges

In the world of emerging businesses, stress and setbacks are inevitable. Whether it's tight deadlines, resource constraints, an investor backing out, the constant pressure to perform, or product feedback that stings, these moments test your leadership and mindset and can push even the most resilient leaders and teams to their limits.

While you can't always control what happens, you *can* control how you respond. This section explores strategies for maintaining focus, motivation, and mental well-being in the face of uncertainty, helping you and your team navigate challenges with confidence and clarity - even when things get tough.

Why Stress Management Matters in Startups

While stress often has a negative reputation, it can also be a source of growth and innovation - if managed effectively. When handled well, setbacks can:

1. **Strengthen Resilience:**
    - Build your ability to bounce back stronger from future challenges.
2. **Foster Innovation:**
    - Inspire creative solutions to overcome obstacles.
3. **Clarify Priorities:**
    - Force you to focus on what truly matters.

The goal isn't to eliminate stress but to manage it in a way that keeps you productive and motivated.

The Hidden Inflection Point Theory

One of the most powerful but least discussed phenomena in startup project management is what experts call the "Hidden Inflection Point." This theory suggests that in nearly every startup project, there is a critical transition point - usually around the 60-70% completion mark - where a significant change in direction or progress is likely to occur and traditional project constraints (scope, time, budget) are overtaken by team dynamics and emotional intelligence as the main drivers of success.

At this inflection point, several predictable challenges emerge:

- Team fatigue begins to set in as the initial excitement wanes
- Scope clarity diminishes as complexities and edge cases emerge
- Motivation naturally decreases as the finish line remains distant but the starting enthusiasm has faded

- Problem-solving becomes more difficult as the "easy wins" have been achieved

This inflection point is “hidden” because it rarely shows up on project timelines or Gantt charts. These points are often not immediately apparent and can be concealed by ongoing project activities, making them challenging to identify until it’s too late to react effectively. Pay close attention to project performance metrics, stakeholder feedback, and emerging project trends to identify potential inflection points.

Examples

- **A project that initially seems to be on track but faces unexpected delays or cost overruns:** This could be an inflection point signaling a need for a reassessment of the project scope or timeline.
- **A project that has reached a plateau in terms of progress or impact:** This could be an inflection point indicating a need for innovation or a shift in strategy to achieve desired outcomes.
- **A project that is facing a change in stakeholder expectations or priorities:** This could be an inflection point requiring a reevaluation of project deliverables and communication strategies.

Proactive adaptation is important. Once an inflection point is identified, be prepared to make strategic adjustments to the project plan, resources, or approach to capitalize on opportunities or mitigate potential risks.

How to Navigate the Hidden Inflection Point

1. **Recognize the Transition**
   - Watch for signs that your team has hit this point - decreased communication, missed deadlines, increased tension, or a general lack of enthusiasm.
   - **Example**: The Airbnb team recognized this phenomenon during their early marketplace development, noting that around month 8 of a 12-month development cycle, their previously high-performing team seemed to lose momentum despite being on schedule.
2. **Shift Your Leadership Approach**
   - Front-load analytical frameworks and strict methodology in the first half of projects
   - Gradually transition to more intuitive, relationship-driven approaches as the project progresses
   - **Example**: Buffer's product team schedules what they call "Project Renewal Days" at the 60% mark of major initiatives, where they step back from tasks to reconnect with the vision and each other.

3. **Implement "Constraint Relaxation"**
    - Temporarily loosen certain formal constraints to allow for creative problem-solving
    - Provide more autonomy during this phase to reignite ownership
    - **Example**: Spotify gives teams what they call "innovation windows" during the late middle phase of projects - short periods when regular metrics and reporting are suspended to encourage breakthrough thinking.
4. **Conduct Energy Audits**
    - Assess which aspects of the project are energizing the team and which are draining them
    - Redistribute tasks based on renewed energy assessments rather than just skill sets
    - **Example**: The team at Basecamp conducts "energy check-ins" at the midpoint of projects, asking team members to rate tasks on an energy scale rather than just a difficulty scale.

Studies of failed startup projects often reveal that they maintained rigid methodological approaches throughout, while successful ones demonstrated this conscious shift in management style precisely at this inflection point. By understanding and preparing for this hidden inflection point, you position your team to maintain momentum through the most challenging phase of project execution.

Benefits of Applying the Theory

- **Improved Project Performance:** By proactively adapting to inflection points, project managers can improve the likelihood of achieving project goals and objectives.
- **Enhanced Stakeholder Satisfaction:** Addressing potential issues or challenges early on can lead to greater stakeholder satisfaction and buy-in.
- **Increased Project Agility:** The ability to identify and respond to inflection points can make projects more agile and adaptable to changing circumstances.

Strategies for Managing Stress and Setbacks

1. Reframe Challenges as Opportunities

Changing how you view a setback can shift your mindset from defeat to determination.

How to Reframe:

- Focus on what you can learn from the experience.
- Ask yourself, "How can this challenge make us better or stronger?"
- Celebrate small wins along the way to keep morale high.

Example: If a product launch is delayed, use the extra time to refine features or improve customer outreach.

2. Prioritize Self-Care

Taking care of yourself isn't a luxury - it's a necessity for effective leadership.

Self-Care Tips:

- **Physical Health:** Exercise regularly, eat well, and get enough sleep.
- **Mental Health:** Practice mindfulness, meditation, or journaling to reduce stress.
- **Boundaries:** Set clear work-life boundaries to avoid burnout.

Pro Tip: Treat self-care like any other critical task on your to-do list - schedule it, commit to it, and don't skip it.

3. Break Problems into Manageable Steps

Big challenges can feel overwhelming. Breaking them into smaller, actionable steps makes them more approachable.

Steps to Take:

1. Define the problem clearly.
2. Brainstorm possible solutions.
3. Prioritize actions based on impact and feasibility.
4. Tackle one step at a time.

Example: If cash flow is tight, start by identifying immediate cost-cutting measures, then explore longer-term revenue opportunities.

4. Lean on Your Team

You don't have to face challenges alone. Engaging your team fosters collaboration and builds collective resilience.

How to Involve Your Team:

- Share the challenge openly and transparently.
- Encourage brainstorming sessions to generate creative solutions.
- Delegate tasks to distribute the workload evenly.

Pro Tip: Use setbacks as an opportunity to strengthen team bonds and demonstrate that challenges are shared, not individual burdens.

5. **Focus on What You Can Control**

Stress often stems from worrying about factors outside your control. Shift your energy to what you can influence.

How to Focus:

- Identify the elements of the situation you can act on.
- Accept and let go of what's beyond your control.
- Stay present and take it one step at a time.

Example: If market conditions shift unexpectedly, focus on adapting your product or marketing strategy rather than dwelling on the external factors.

6. **Build a Support Network**

Surrounding yourself with mentors, advisors, or peers provides perspective and encouragement.

How to Leverage Your Network:

- Share your challenges and seek advice from trusted mentors.
- Join startup communities or networking groups for shared experiences.
- Don't hesitate to ask for help when you need it.

Pro Tip: Sometimes, just talking through a problem with someone outside your immediate team can spark new ideas or solutions.

Common Pitfalls in Managing Stress and Setbacks

1. **Ignoring the Problem:**
   - **The Problem:** Hoping a challenge will resolve itself can lead to bigger issues.
   - **The Solution:** Address problems proactively and break them into manageable steps.
2. **Trying to Do It All Yourself:**
   - **The Problem:** Taking on too much can lead to burnout.
   - **The Solution:** Delegate tasks and lean on your team for support.
3. **Overreacting to Setbacks:**
   - **The Problem:** Small challenges feel insurmountable when you're stressed.
   - **The Solution:** Take a step back, reassess, and focus on actionable steps.

4. **Missing the Hidden Inflection Point:**
   - **The Problem:** Failing to recognize when your project needs a shift in leadership approach.
   - **The Solution:** Proactively plan for the 60-70% mark of your project with relationship-focused strategies.

Example: Managing a Major Setback

Case Study: Overcoming a Funding Shortfall

Imagine a startup that planned to launch a product with a recent funding round, but an investor unexpectedly pulls out, leaving the team short on cash.

1. **Reframe the Challenge:**
   - The team views the setback as an opportunity to streamline their operations and validate the product with fewer resources.
2. **Prioritize Actions:**
   - They reduce spending on non-essential areas and focus efforts on building an MVP to attract new investors.
3. **Engage the Team:**
   - The founder holds a transparent meeting to discuss the situation and brainstorm solutions, fostering collaboration and commitment.
4. **Leverage Support:**
   - They reach out to advisors for introductions to potential investors and insights on managing lean budgets.
5. **Navigate the Hidden Inflection Point:**
   - Recognizing that the team was already at the 65% mark of their development timeline, the founder shifts from metrics-focused management to a more supportive, relationship-driven approach, scheduling one-on-one check-ins and creating space for creative problem-solving.

Outcome: The startup successfully builds an MVP, secures new funding, and moves forward with a leaner, more focused strategy.

Avoiding The Pitfalls

Pitfall	Better Strategy
Ignoring stress	Acknowledge it - then manage it proactively
Trying to do it all alone	Ask for help. Delegate. Share the burden.
Overreacting to setbacks	Pause. Reframe. Respond strategically.
Skipping self-care	Treat rest like fuel for performance
Taking it personally	Learn from it - but don't make it about your worth

The Takeaway

Stress and setbacks are inevitable in emerging businesses, but they don't have to derail your journey. What matters is how you respond. By reframing challenges, focusing on what you can control, and leaning on your team and support network, you can navigate uncertainty with resilience and confidence. Remember, every obstacle is an opportunity to learn, grow, and come back stronger.

Every great founder has a few scars - and a story about how they kept going when it got hard.

Understanding the Hidden Inflection Point Theory gives you a powerful advantage - the ability to anticipate and prepare for the natural ebb in project momentum that most teams face. By proactively shifting your leadership style from process-driven to people-driven at this critical juncture, you can help your team maintain energy and creativity through the most challenging phase of project execution.

9

Project Reporting and Communication

Effective communication is the lifeblood of successful project management, and reporting is its most structured form. Yet, reporting often becomes an afterthought or an overwhelming burden for many startups and emerging ventures. This chapter explores how to strike the right balance - creating project reports that provide real value without consuming excessive time or resources.

1. Right-Sizing Your Reporting Approach

Not every project requires the same level of reporting rigor. A startup launching its MVP needs a different approach than a team managing a multi-million dollar infrastructure project. The key is to match your reporting structure to your project's complexity, stakeholder needs, and organizational culture.

The Reporting Spectrum: From Lightweight to Comprehensive

Project reporting exists on a spectrum, and understanding where your project falls is the first step in creating an effective approach:

Reporting Level	Best For	Key Characteristics
Lightweight	Small teams, internal projects, low-stakes initiatives	Informal updates, minimal documentation, focus on quick alignment
Moderate	Cross-functional projects, projects with external stakeholders	Structured but concise reports, balance of metrics and narrative
Comprehensive	High-visibility projects, regulated environments, complex initiatives	Detailed documentation, formal approval processes, extensive metrics

The mistake many startups make is assuming they need comprehensive reporting when a lightweight approach would be more effective. The goal is to provide just enough information to drive decisions and keep stakeholders aligned - no more, no less.

Factors That Determine Your Reporting Needs

To right-size your reporting approach, consider these key factors:

1. **Project Complexity**:
   - Simple projects with few moving parts require simpler reporting.
   - Complex projects with multiple workstreams need more detailed tracking.
2. **Stakeholder Expectations**:
   - Some stakeholders (especially external ones) may require formal reporting.
   - Internal teams might prefer quick, informal updates.
3. **Risk Level**:
   - High-risk projects benefit from more frequent and detailed reporting.
   - Low-risk initiatives can use streamlined approaches.
4. **Team Size and Distribution**:
   - Larger, distributed teams typically need more structured communication.
   - Small, co-located teams can often rely on more informal methods.
5. **Regulatory Requirements**:
   - Some industries have mandatory reporting requirements that must be met.

Right-Sizing in Action: A Real-World Example

Consider two different startups, both managing important projects:

Startup A: Mobile App Development

- Small, co-located team of 5 people
- Building an MVP for release in 3 months
- Bootstrap funded with the founding team as primary stakeholders

Appropriate Reporting Approach:

- Weekly stand-up meetings with key metrics tracked on a simple dashboard
- Brief written summary shared in Slack each Friday
- Monthly retrospective to discuss progress and challenges

Startup B: Healthcare Software Implementation

- Team of 15 across multiple locations
- Implementing software for healthcare providers over 6 months
- Funded by investors and subject to healthcare regulations

Appropriate Reporting Approach:

- Formal weekly status reports with defined metrics
- Bi-weekly stakeholder review meetings with presentation decks
- Detailed monthly progress reports for investors and regulatory compliance
- Documented risk assessments and mitigation plans

The difference isn't about one approach being "better" than the other - it's about matching the reporting style to the project's needs. Startup A would be hindered by excessive reporting, while Startup B would face significant risks without structured communication.

How to Determine the Right Level of Reporting

1. **Start with a Stakeholder Analysis**:
   - Who needs information about your project?
   - What decisions do they need to make based on that information?
   - How do they prefer to receive updates?
2. **Assess Your Project's Risk Profile**:
   - What could go wrong, and what would the impact be?
   - Higher-risk areas warrant more detailed tracking and reporting.
3. **Consider Your Resources**:
   - Be realistic about the time and tools available for reporting.
   - Choose an approach that's sustainable throughout the project lifecycle.
4. **Begin Simple and Adjust**:
   - Start with the minimum viable reporting structure.
   - Add elements only as needed based on feedback and project evolution.

Common Reporting Anti-Patterns to Avoid

1. **Report Inflation**:
   - Creating lengthy reports filled with details no one needs or reads.
   - **Solution**: Focus on actionable insights rather than exhaustive data.

2. **One-Size-Fits-All Reporting**:
    - Using the same reporting structure for every project regardless of complexity.
    - **Solution**: Tailor your approach based on the project's specific needs.
3. **Reporting as a Checkbox**:
    - Creating reports to "check the box" rather than to drive decisions.
    - **Solution**: Ensure every report has a clear purpose and audience.
4. **All Numbers, No Context**:
    - Sharing metrics without explaining what they mean for the project.
    - **Solution**: Include narrative that turns data into insights.

The Takeaway

The most effective reporting approach is one that provides the right information to the right people at the right time - no more, no less. By right-sizing your reporting to match your project's complexity and stakeholder needs, you can ensure communication supports rather than hinders your progress.

In the next section, we'll explore how to craft reports for different stakeholders, ensuring each audience gets the information they need in a format that works for them.

2. Crafting Reports for Different Stakeholders

Different stakeholders need different information. What's valuable to your engineering team might be meaningless to your investors, and what your CEO needs to know differs from what your customers care about. Effective reporting means tailoring your communication to each audience's specific needs, priorities, and preferences.

Understanding Stakeholder Information Needs

The first step in crafting effective reports is understanding what each stakeholder group actually needs to know. Here's a framework for common stakeholder groups in startups:

Stakeholder	Primary Concerns	Information Needs
Executives/Founders	Strategic alignment, ROI, risks	High-level progress, key metrics, major blockers, strategic implications
Team Members	Day-to-day tasks, dependencies	Detailed task breakdowns, priorities, immediate next steps
Investors	Financial returns, milestones, market traction	Progress against key metrics, market validation, resource utilization
Customers	Product delivery, value realization	Feature updates, timelines that affect them, how to provide feedback
External Partners	Interface points, dependencies	Specific deliverables that affect their work, key dates, collaboration needs

The Report Pyramid: Different Levels of Detail

Think of your reporting approach as a pyramid:

1. **Executive Summary (Top)**:
    - Brief overview highlighting key achievements, challenges, and decisions needed
    - Ideal for executives and investors who need quick insights
2. **Core Metrics and Status (Middle)**:
    - Key performance indicators, milestone progress, and risk assessments
    - Relevant for managers and stakeholders with direct involvement
3. **Detailed Updates (Bottom)**:
    - Task-level progress, specific technical details, and comprehensive data

- Appropriate for team members and those directly working on the project

Not every report needs all three levels. The key is matching the level of detail to the stakeholder's needs and role in the project.

Tailoring Content to Different Stakeholders

Here's how to customize your reporting for key stakeholder groups:

For Executives and Investors:

- Focus on business outcomes and strategic impact
- Highlight progress against key milestones and metrics
- Provide clear, actionable insights rather than raw data
- Be transparent about risks and challenges, but always pair them with mitigating actions

Sample Format: One-page executive dashboard with 3-5 key metrics, major accomplishments, critical risks, and decisions needed.

For Project Team Members:

- Emphasize task-level details and immediate priorities
- Show dependencies and how individual work contributes to project goals
- Provide visibility into upcoming work and potential roadblocks
- Include technical details relevant to implementation

Sample Format: Task board with current status, next steps, and blockers identified during standups or team meetings.

For Customers or End Users:

- Focus on features, benefits, and delivery timelines
- Use accessible language, avoiding internal jargon
- Highlight opportunities for feedback or involvement
- Communicate changes that will impact their experience

Sample Format: Product roadmap visualization with upcoming features and release dates.

For External Partners and Vendors:

- Concentrate on interface points and shared dependencies
- Clearly articulate expectations and deadlines
- Provide context on how their work fits into the broader project
- Highlight collaboration opportunities and requirements

Sample Format: Timeline showing integration points, delivery dates, and collaboration sessions.

Communication Channels Matter

The channel you choose for reporting is just as important as the content itself:

Channel	Best For	Considerations
Email Reports	Formal updates, distributed teams, creating a record	Can get buried, may not prompt discussion
In-Person/Video Meetings	Complex updates, discussions requiring input, building relationships	Time-intensive, scheduling challenges
Project Management Tools	Day-to-day tracking, team collaboration, visualizing progress	Requires everyone to use the system
Dashboards	Real-time metrics, self-service information, transparent progress tracking	Initial setup time, needs regular maintenance
Chat/Messaging	Quick updates, immediate questions, informal team communication	Can become noisy, important info gets lost

Choose channels that match both your stakeholders' preferences and the type of information being shared.

The Art of Translation

One of the most valuable skills in project reporting is the ability to "translate" information across different domains:

- Technical details → Business implications
- Raw data → Actionable insights
- Complex concepts → Accessible explanations
- Individual tasks → Overall project progress

This translation ensures each stakeholder not only receives information but understands what it means for them and what actions (if any) they should take as a result.

A Real-World Example: Stakeholder-Specific Reporting

Let's look at how a startup launching a new e-commerce platform might tailor its reporting:

For the CEO and Board:

- Monthly report focusing on the platform's readiness for launch
- Key metrics: development progress (%), burn rate, projected timeline vs. original
- High-level risks and mitigation strategies
- Strategic decisions needed about launch timing and marketing approach

For the Development Team:

- Weekly sprint updates in the project management tool
- Daily standup focusing on immediate tasks, blockers, and dependencies
- Technical debt dashboard showing issues to address
- Quality metrics including bug counts and test coverage

For Early Access Customers:

- Bi-weekly email highlighting new features they can test
- Simple feedback mechanism for reporting issues or requesting features
- Clear timeline showing when they can expect specific functionality
- Frequently asked questions and troubleshooting guides

By tailoring both the content and the delivery method to each stakeholder group, the startup ensures everyone has the information they need without overwhelming anyone with irrelevant details.

3. Establishing Effective Reporting Cadences

Determining when to report is just as important as deciding what to report. Too frequent, and reporting becomes a burden that takes time away from actual work. Too infrequent, and stakeholders lose visibility, potentially missing critical issues until it's too late to address them effectively.

The right reporting cadence creates a rhythm that keeps everyone informed without overwhelming your team with administrative overhead.

Matching Cadence to Project Phases

Different phases of your project may require different reporting frequencies:

Project Phase	Recommended Cadence	Rationale
Initiation	Weekly or bi-weekly	Establish alignment and clarity as the project takes shape
Planning	As milestones are reached	Focus on key planning deliverables rather than time-based updates
Execution (normal progress)	Weekly or bi-weekly	Regular cadence to track progress and address issues
Execution (critical periods)	Daily or twice weekly	Increased visibility during high-risk or intense work periods
Nearing Launch	Twice weekly or daily	Close monitoring as you approach key deadlines
Post-Launch/ Closeout	Weekly, then transitioning to monthly	Track immediate issues, then shift to monitoring long-term impact

The key is flexibility - your reporting cadence should adapt to the project's current needs and risk level.

Establishing Multi-Level Reporting Rhythms

Most projects benefit from a multi-level reporting approach with different frequencies for different stakeholder groups:

Daily: Brief team check-ins (e.g., standups)

- Focus on: Immediate tasks, blockers, coordination needs
- Format: 15-minute meetings or chat updates
- Participants: Core team members

Weekly: Team progress and coordination

- Focus on: Sprint progress, upcoming work, issue resolution
- Format: 30-60 minute meetings with visual task boards
- Participants: Project team and immediate managers

Bi-weekly or Monthly: Executive/stakeholder updates

- Focus on: Milestone progress, key metrics, strategic decisions
- Format: Structured report or presentation with Q&A
- Participants: Leadership, key stakeholders, sponsor representatives

Quarterly: Strategic reviews

- Focus on: Overall progress, market alignment, major adjustments
- Format: Comprehensive presentation with discussion
- Participants: Executives, investors, strategic partners

This multi-level approach ensures each group gets the information they need at a frequency that makes sense for their involvement in the project.

Avoiding "Reporting Fatigue"

Even with the best intentions, reporting can become a burden that drains team productivity and morale. Here are strategies to prevent reporting fatigue:

1. **Automate Where Possible**:
    - Use project management tools that generate reports automatically
    - Set up dashboards that update in real-time
    - Create templates that can be quickly populated with current data
2. **Implement Progressive Elaboration**:
    - Start with minimal reporting and add elements only as needed
    - Regularly review and eliminate reports that don't drive decisions
    - Ask stakeholders what information they actually use
3. **Combine Reporting Activities**:
    - Use team meetings to gather status updates
    - Create one core report that can be filtered for different audiences
    - Align reporting with natural project milestones
4. **Make Reporting Valuable for the Team**:
    - Ensure reporting helps the team identify and resolve issues
    - Use reporting sessions to celebrate progress and wins
    - Create two-way communication, not just one-way status updates

The Stakeholder Communication Plan

To formalize your approach, create a simple stakeholder communication plan that captures who needs what information and when:

1. **List all stakeholders** or stakeholder groups
2. **Identify their information needs** (refer to the previous section)
3. **Determine appropriate frequency** for each type of communication
4. **Choose the most effective channel** for each stakeholder
5. **Assign responsibility** for creating and delivering each report
6. **Get stakeholder agreement** on the plan before implementing

Most importantly, schedule check-ins to evaluate whether your reporting cadence is working. Ask stakeholders if they're getting the information they need, and ask your team if the reporting process is sustainable.

Real-World Example: Adjusting Cadence Mid-Project

A software startup was providing weekly detailed reports to all stakeholders during their product development. They noticed:

1. The reports were taking a full day to prepare each week
2. Few stakeholders were reading the detailed sections
3. Team members were frustrated by the reporting overhead

Their **solution**:

- Replaced the weekly detailed report with a real-time dashboard in Asana
- Shifted to bi-weekly executive updates with only key metrics and blockers
- Implemented daily 10-minute standups focused exclusively on blockers
- Created a monthly in-depth review for those who wanted details

The **result**: Reporting time decreased by 75%, stakeholder satisfaction increased, and the team had more time to focus on actual product development.

4. Tools and Templates for Efficient Reporting

The right tools and templates can transform reporting from a tedious chore into a streamlined process that adds real value. In this section, we'll explore practical solutions to make your reporting more efficient, engaging, and effective - without breaking your startup's budget or overwhelming your team.

Essential Reporting Tools for Startups

Modern project management tools offer powerful reporting capabilities that can save countless hours of manual work. Here are some options suitable for different startup needs and budgets:

Tool	Best For	Price	Key Reporting Features	URL
Trello	Visual task tracking, simple projects	Free tier available; Business Class from $10/user/month	Customizable views, due date tracking, power-up integrations	trello.com
Asana	Cross-functional projects, workflow management	Free tier available; Premium from $10.99/user/month	Timeline views, dashboard reporting, automated status updates	asana.com
ClickUp	Comprehensive project management, customizable workflows	Free tier available; Unlimited from $7/user/month	Custom dashboards, sprint reporting, goal tracking	clickup.com
Monday.com	Visual project tracking, team collaboration	Basic from $8/user/month	Customizable dashboards, automated reporting, client-friendly views	monday.com
Notion	Documentation and project management hybrid	Free tier available; Team from $8/user/month	Database views, linked documents, customizable templates	notion.so
Google Sheets	Custom reporting, data visualization	Free	Custom formulas, conditional formatting, chart creation	sheets.google.com
Microsoft Excel	Teams who need customization options and are already familiar with Excel	Included with Microsoft 365 from $6/user/month; One-time purchase options available	Budget tracking, resource allocation, and creating Gantt charts without a learning curve	microsoft.com/microsoft-365/excel

The best tool is one that your team will actually use. Consider these factors when choosing:

- Ease of adoption for your team
- Integration with your existing tools
- Scalability as your startup grows
- Value for your specific reporting needs

Templates

While dedicated reporting tools have their place, reporting platforms that are powerful and flexible are the key criteria for growing businesses. Ones with the following features:

- **Dynamic dashboards** that update automatically as data changes
- **Instant data analysis** without coding knowledge
- **Custom charts** tailored to visualize exactly what matters
- **Conditional formatting** to highlight exceptions and trends
- **Data connections** to pull information from other sources
- **Export options** to share in various formats (PDF, image, web)

As a project manager in a dynamic business, that last thing you want to be doing is spending significant time creating reports and charts. Automate as much of the reporting side of your role as possible.

Reporting templates worth implementing:

1. **Project Status Dashboard** - This could take several forms such as a detailed Weekly Progress Report for team members or immediate management, or a one-page overview with RAG status indicators
2. **Resource Utilization Report** - Visual tracking of team capacity and allocation
3. **Milestone Tracking Chart** - Timeline visualization with completion status
4. **Budget vs. Actual Report** - Financial performance with variance analysis
5. **Risk Register** - Heat map visualization of project risks and mitigation status

Essential Templates for Project Reporting

Templates save time and ensure consistency. Here are essential templates for startup project reporting. Consider these as samples only that have worked in other businesses. They may need to be, and should be, adjusted to suit your particular organisation's specific requirements.

1. Weekly Status Report Template

A concise overview of progress, issues, and upcoming work.

WEEKLY STATUS REPORT		
PROJECT:	[Project Name]	
REPORTING PERIOD:	[Date Range]	
PROGRESS SUMMARY		
TASKS COMPLETED	[Major tasks/milestones completed]	
TASKS IN PROGRESS	[Key tasks currently underway]	
TASKS BLOCKED	[Items needing attention]	
KEY METRICS	**CURRENT VALUE**	**TARGET VALUE**
[Metric 1]		
[Metric 2]		
[Metric 3]		
RISKS & ISSUES	**IMPACT**	**MITIGATION PLAN**
[Risk / Issue 1]		
[Risk / Issue 2]		
[Risk / Issue 2]		
NEXT WEEK'S FOCUS		
[Priority Task 1]		
[Priority Task 2]		
DECISIONS NEEDED		**REQ'D BY DATE**
[Decision 1]		
[Decision 2]		
[Decision 3]		
[Decision 4]		

2. Executive Dashboard Template

A visual one-pager for high-level stakeholders.

Metric	Status	Trend	Notes
Schedule	On Track	Improving	Ahead on milestone 3
Budget	Monitor	→ Stable	85% of budget used, 80% of work complete
Scope	On Track	→ Stable	No significant scope changes
Resources	At Risk	Declining	Designer capacity constrained
Quality	Monitor	Improving	Bug count decreasing after recent testing

3. Project Milestone Tracker

Track progress against key project milestones.

Milestone	Due Date	Status	Owner	% Complete	Notes
MVP Design	15-Mar	Complete ✓	Sarah	100%	Approved on 12-Mar
Backend Development	30-Apr	At Risk	Miguel	65%	API integration delayed
User Testing	15-May	Not Started	Taylor	0%	Dependent on backend completion
Marketing Materials	01-Jun	On Track	Jordan	40%	Social media templates complete

4. Risk Register Template

Track and manage project risks:

Risk	Prob. (1-5)	Impact (1-5)	Risk Score	Owner	Mitigation Strategy	Status
API integration delays	4	5	20	Miguel	Begin integration with mock data	Active
User adoption uncertainty	3	4	12	Jordan	Plan beta testing with key users	Monitoring
Budget constraints for marketing	2	3	6	Alex	Prioritize organic channels	Resolved

Visualizing Project Data Effectively

The way you present information significantly impacts how it's understood and acted upon. Here are techniques for visualizing project data effectively:

1. Progress Visualization

- **Burndown/Burnup Charts**: Show work completed versus time

- **Gantt Charts**: Visualize timeline and dependencies
- **Kanban Boards**: Display task status across workflow stages

2. Metric Visualization

- **Speedometers/Gauges**: Show performance against targets
- **Trend Lines**: Display metrics over time to identify patterns
- **Heat Maps**: Highlight problem areas or concentrations

3. Status Communication

- **Traffic Lights**: Use red/yellow/green for quick status overview
- **Progress Bars**: Show completion percentage for major workstreams
- **Milestone Charts**: Highlight key achievements and upcoming targets

Tools for Creating Visual Reports

These tools can help you create compelling visual reports without design expertise:

- **Google Data Studio**: Create interactive dashboards (Free) - datastudio.google.com
- **Canva**: Design professional-looking reports (Free tier available) - canva.com
- **Miro**: Collaborative visual workspace (Free tier available) - miro.com
- **Power BI**: Powerful data visualization (Free desktop version) - powerbi.microsoft.com
- **Tableau Public**: Sophisticated data visualization (Free) - tableau.com
- **Microsoft Excel**: With free or commercially available templates

Building Your Own Reporting System

For most startups, a combination of tools works best for project reporting:

1. **A central project management tool** (e.g., Asana, ClickUp) for day-to-day tracking
2. **A visualization tool** (e.g., Google Data Studio) for dashboards and metrics
3. **A communication platform** (e.g., Slack, email) for distributing updates
4. **Templates** for consistent reporting formats

Start simple and add complexity only as needed. A Google Sheet or Microsoft Excel dashboard is often sufficient for early-stage startups, while more established teams might benefit from dedicated reporting tools.

Automating Your Reporting

Automation can drastically reduce the time spent on reporting:

1. **Use Integrations**:
   - Connect your project management tool to Slack for automated updates
   - Set up email digests from your PM tool to stakeholders
   - Use Zapier to connect tools that don't natively integrate
2. **Create Dashboards**:
   - Build dashboards that pull live data from your project tools
   - Share dashboard links instead of creating manual reports
   - Set up automated refresh schedules
3. **Implement Templates with Formulas**:
   - Create spreadsheets that automatically calculate key metrics
   - Use document templates with placeholders for easy updates
   - Build presentation decks with linked data sources

Real-World Example: A Startup's Reporting Evolution

A fintech startup evolved its reporting approach as it grew:

Stage 1: Pre-seed (3 people)

- Daily in-person check-ins
- Shared Google Sheet for tracking tasks and milestones
- Weekly email to their two angel investors

Stage 2: Seed Funding (10 people)

- Implemented Asana for task tracking
- Created a Google Data Studio dashboard for key metrics
- Bi-weekly stakeholder meeting with a simple presentation

Stage 3: Series A (25 people)

- Upgraded to ClickUp for comprehensive project management
- Automated Slack updates for daily progress
- Built custom reporting dashboards for different departments
- Monthly formal investor reports

The key to their success: They adapted their reporting approach to match their growth stage rather than implementing overly complex systems too early.

The Takeaway

Effective reporting doesn't require expensive tools or excessive time - it requires thoughtful selection of the right tools and templates for your specific needs. By starting with simple, focused reports and building as your startup grows, you can create a reporting system that enhances decision-making without draining resources.

Remember: The best reporting system is one that provides valuable insights with minimal effort, keeping your team focused on what matters most - building and growing your business.

10

Real-World Startup Project Stories

1. Successes and Failures

Lessons from the Trenches of Startup Project Management

Startups don't grow in straight lines. They zig, they zag, and sometimes they crash. Behind every success headline is a messy middle. Behind every failure is a lesson worth learning.

In this section, we explore real-world case studies - some inspiring, others cautionary. Whether the result was triumph, collapse, or reinvention, each of these startup stories reveals something powerful about the realities of project management, decision-making, and adapting under pressure. Each story illustrates practical lessons that can guide you in your own projects, whether you're navigating your first launch or scaling for growth.

Success Stories: What Went Right

1. Case Study: Dropbox - Starting Simple to Scale Big

Industry: Technology

The Project: Cloud storage MVP

Dropbox's story exemplifies smart project management in a crowded file-syncing market. Founders Drew Houston and Arash Ferdowsi prioritized simplicity over feature bloat, releasing a three-minute demo video in 2007 that showcased intuitive cloud storage—generating 75,000 sign-ups overnight from a modest TechCrunch audience and validating demand without building the full product first. This lean MVP approach, paired with early beta feedback loops to refine usability, fueled rapid iteration and growth into a platform now serving hundreds of millions worldwide. The lesson for startups: Start with a simple vision, test it leanly with users, then scale confidently.

Key Project Management Wins:

- **Clear Vision:** Dropbox concentrated on making file syncing simple and intuitive rather than packing in complexity.
- **MVP Focus:** The demo video served as a pre-launch validation tool, helping the team test demand before investing heavily in the full product.
- **Customer Feedback Loop:** Early beta users provided feedback that helped Dropbox refine the product and keep the experience straightforward.

Outcome:
By starting with a focused concept, validating demand early, and iterating from user feedback, Dropbox grew into a global cloud-storage and collaboration company used by hundreds of millions of people.

Lesson:
Start simple. Validate demand. Then build.

2. Case Study: Canva - Growth Through Listening

Industry: Design tools

The Project: An easy-to-use platform for non-designers

Canva's ascent from a 2013 beta waitlist to design dominance shows the power of customer-led project management. Founders Melanie Perkins, Cliff Obrecht, and Cameron Adams targeted non-designers like bloggers and influencers for early access, iterating swiftly on feedback to perfect an intuitive drag-and-drop platform. This agile, collaborative approach—rooted in values like simplifying complexity—drove 300,000 users in year one and explosive growth to 220M+ monthly actives today. Lesson for startups: Listen to early users as your co-product managers, adapt relentlessly, and scale with simplicity.

Why It Worked:

- **Customer-Centric Iteration:** The team granted beta access to vocal early adopters, using their feedback to refine UX and features rapidly.
- **Collaborative Culture:** Strong internal values like "empower others" and "make complex things simple" fostered cohesion for fast scaling.
- **Agility:** Quick pivots based on user input ensured high satisfaction and organic growth.

Outcome:
Canva now boasts over 220 million monthly active users across 190 countries, serving individuals, businesses, and educators as a go-to design platform.

Lesson:
Stay close to your customers and adapt fast. They're your best product managers - if you listen.

Failure Stories: What Went Wrong

1. Case Study: Quibi - Big Vision, No Validation

Industry: Streaming media

The Project: Premium short-form video app

Quibi's $1.75 billion flameout after a 6-month April-October 2020 run warns against unchecked ambition in project management. Backed by Jeffrey Katzenberg and Meg Whitman, it bet big on premium short-form mobile video without MVP testing, ignoring free competitors like TikTok and misjudging demand. COVID obliterated its commuting use case while gimmicks like non-shareable portrait content repelled users, proving even star power can't fix poor validation. Lesson for startups: Test rigorously before burning cash—market fit trumps funding every time.

Why It Failed:

- **Weak Product-Market Fit:** TFounders assumed demand for premium mobile-only shorts without testing; free alternatives like TikTok already dominated.
- **Wrong Timing:** COVID lockdowns killed the "on-the-go" use case, pushing users to long-form home viewing on Netflix and YouTube.
- **Overbuilt Too Early:** No MVP or beta—hundreds of millions spent on unproven content and features like portrait/landscape "Turnstyle" that users ignored.

Outcome:
Quibi shut down in October 2020 after 6 months, writing off $1.75 billion and becoming a textbook startup cautionary tale.

Lesson:
Resources don't equal readiness. Validate assumptions with real data before scaling—timing and adaptability matter as much as vision.

2. Case Study: Jawbone - Failing to Align Execution with Strategy

Industry: Consumer hardware

The Project: Fitness trackers and audio products

Jawbone pioneered sleek wearables—from Bluetooth headsets to Jambox speakers and UP fitness trackers—raising nearly $1 billion and hitting a $3 billion valuation, yet faltered by 2017 due to misaligned execution. Spreading thin across hardware, software, and pivots drained resources, while persistent quality woes (faulty bands, sync failures) eroded trust amid Fitbit/ Apple competition; cash mismanagement and Fitbit lawsuits sealed its fate. Lesson for hardware startups: Innovation dazzles, but disciplined focus, reliable delivery, and sustainable finances win.

Why It Failed:

- **Too Many Fronts:** Juggling hardware, software, and new product lines stretched resources.
- **Quality Issues:** Frequent bugs and product flaws damaged credibility.
- **Cash Flow Mismanagement:** Burnout on capital with low ROI.

Outcome:
Jawbone filed for bankruptcy in 2017, leaving behind valuable lessons for startups in hardware development.

Lesson:
Execution and focus matter more than innovation alone. Even great ideas can fail without disciplined management.

Pivot-to-Win: From Failure to Success

1. Case Study: Slack - Messaging from the Ashes

Industry: Enterprise software

The Project: Internal tool turned global platform

Slack's pivot from doomed game Glitch to enterprise titan proves failure's hidden upside in project management. When Tiny Speck shut down its 2011 multiplayer flop in 2012, founders like Stewart Butterfield repurposed their slick internal chat tool—solving real team comms pain better than email—into Slack's 2013 beta, fueling viral growth via freemium and integrations. Relentless iteration turned it into a $27B+ powerhouse acquired by Salesforce. Lesson for startups: Watch your tools closely; pivots from flops often unlock the real market fit.

Why It Worked:

- **Quick Pivot:** In late 2012, the team shifted focus entirely when Glitch wound down, rebranding the company as Slack.
- **Real Pain Point:** Their internal chat tool addressed fragmented team communication better than email or IRC.
- **Obsessive Iteration:** Relentless polishing—search, integrations, freemium—drove viral adoption.

The Pivot:
The team bet on the communication sidecar, launching a public beta in August 2013, later rebranding as Slack.

Outcome:
Slack revolutionized workplace messaging, growing to millions of users and a $27B+ valuation before its 2021 Salesforce acquisition.

Lesson:
Failed projects can birth unexpected winners—if you spot the signal in the noise and pivot boldly on data.

Common Threads and Takeaways

1. **Start Lean:**
   Great startups begin with MVPs, not masterplans. Focus on an MVP to test your idea before investing heavily.
2. **Validate Early:**
   Test assumptions with real user feedback and adapt accordingly before you commit big.
3. **Stay Agile:**
   Be ready to pivot when the data demands it.
4. **Focus on Execution:**
   A brilliant idea still needs disciplined execution to succeed.
5. **Learn from Setbacks:**
   The best lessons often come from what didn't work. Setbacks are opportunities to learn and improve.

The Takeaway

Successes inspire. Failures teach. Pivots transform. Every founder - whether they built Dropbox, launched Slack, or watched a promising idea unravel - has something to teach us about startup project management.

Your own story will include highs, lows, and hard calls. The key isn't to avoid failure - it's to learn from it, fast.

Every successful startup is built on a pile of learnings. The faster you collect them, the stronger you grow.

In the next section, we'll dig deeper into the **lessons behind the stories** - unpacking what these founders did well (and poorly) when it came to execution, leadership, and strategic decision-making.

2. Lessons Learned

What Worked, What Didn't, and How to Apply It

Every startup project is an experiment. Some succeed spectacularly. Others fail - sometimes quietly, sometimes with fireworks. But all of them leave behind something valuable: lessons.

This section distills those lessons into clear, actionable insights drawn from the real-world startup stories in the previous section. Whether you're launching your first product or managing a pivot, these takeaways can help you steer your next project with greater clarity and confidence.

Lesson 1: Validate Before You Build

Why it matters:
Startups often waste time building features no one needs. Early validation helps you avoid this trap and focus on solving real problems.

What Worked:

- *Dropbox* used a simple explainer video to validate interest before building the product.
- *Airbnb* tested their concept by renting out their own apartment.

What Didn't Work:

- *Quibi* launched a heavily funded platform without validating user demand.

How to Apply It:

- Build an MVP to test the core idea.
- Talk to users early and often.
- Make decisions based on real feedback, not assumptions.

Lesson 2: Focus on One Problem at a Time

Why it matters:
Trying to do everything at once spreads your team (and budget) too thin. Focus creates momentum.

What Worked:

- *Canva* focused on making design accessible for non-designers - and did it exceptionally well.
- *Slack* solved a single pain point: team communication.

What Didn't Work:

- *Jawbone* diluted their focus by launching too many products too quickly.

How to Apply It:

- Define one core problem to solve.
- Say no to feature creep.
- Revisit your focus regularly as your product grows.

Lesson 3: Adapt Quickly to Feedback

Why it matters:
The faster you respond to real-world feedback, the more likely you are to find product-market fit.

What Worked:

- *Slack* pivoted from a failing game to a world-changing communication tool.
- *Airbnb* iterated their model and platform multiple times before finding traction.

What Didn't Work:

- *Theranos* ignored warning signs and feedback from regulators, users, and internal teams.

How to Apply It:

- Set up fast feedback loops.
- View pivots as growth, not failure.
- Create space for your team to raise concerns or share new ideas.

Lesson 4: Execution is Everything

Why it matters:
A great idea means nothing without disciplined follow-through.

What Worked:

- *Stripe* made payment processing seamless by obsessing over the details.
- *Dropbox* built a product that "just worked" right from the first click.

What Didn't Work:

- *Theranos* promised more than they could deliver - and collapsed under the weight of that gap.

How to Apply It:

- Break work into clear milestones.
- Define what “done” looks like before you start.
- Monitor quality and user experience continuously.

Lesson 5: Build a Resilient Culture

Why it matters:
Startups don’t succeed on the back of hustle alone. Team culture sustains momentum through tough times.

What Worked:

- *Atlassian* prioritized transparency and autonomy to build a strong team foundation.
- *Airbnb* weathered early challenges by rallying around shared values.

What Didn’t Work:

- Startups that ignore morale, especially during high-pressure periods, risk burnout and turnover.

How to Apply It:

- Talk openly about challenges.
- Celebrate small wins.
- Encourage cross-functional collaboration and shared ownership.

Lesson 6: Stay Close to the Numbers

Why it matters:
Even promising startups can run out of money. Financial discipline is non-negotiable.

What Worked:

- *Airbnb’s cereal box campaign* during their early cash crunch showed scrappy financial creativity.
- *Stripe* grew sustainably by aligning pricing and growth strategy.

What Didn’t Work:

- *Jawbone* struggled with cash flow and financial forecasting - eventually running out of time.

How to Apply It:

- Track burn rate, runway, and ROI regularly.
- Invest where you see traction, not just where it’s exciting.

- Build financial plans for different scenarios (best case, expected case, worst case).

Lesson 7: Timing is Crucial

Why it matters:
A brilliant idea launched at the wrong time often fails. The best startups hit when the market is ready - and when they're prepared to deliver.

What Worked:

- *Instagram* launched as smartphone cameras and mobile social networks were exploding.
- *Slack* grew with the rise of remote and distributed teams.

What Didn't Work:

- *Quibi* launched during a global pandemic with a mobile-only platform, missing the moment.

How to Apply It:

- Monitor market shifts before and during launch.
- Be prepared to pause - or accelerate - based on timing.
- Reassess timing regularly as part of your strategic planning.

The Takeaway

Emerging businesses don't just require good ideas - they require smart execution, real feedback, flexibility, and endurance. The stories of Dropbox, Canva, Slack, Airbnb, and others prove that success is rarely instant or linear. It's a process of testing, adjusting, and learning at every step.

Apply these lessons to your own journey, and you won't just avoid common pitfalls - you'll be better prepared to create something truly impactful.

In the next section, we'll zoom in on a hands-on example - a small business case study that shows how project management principles can drive success even in scrappy, low-budget environments. Let's bring these insights to life.

3. A Small Business Example

How One Solopreneur Used Project Management to Launch a Product

Not every startup story begins with a venture capital check or a team of developers. Sometimes, it begins at a kitchen table, with a notebook, a dream - and a plan.

Even in the smallest of businesses, effective project management can be the difference between a great idea that fizzles out and one that thrives. This story follows a solopreneur who leveraged simple, practical project management techniques to successfully launch a product, proving that you don't need a large team or a big budget to achieve big results.

Meet the Solopreneur

Sarah is a freelance designer with a passion for sustainable living. She decided to create a line of eco-friendly planners designed to help people live more intentionally. With no team, a limited budget, and only evenings and weekends to work on her project, Sarah knew she needed a clear plan to bring her idea to life.

The Challenge

Sarah faced several hurdles typical of a small business:

1. **Time Constraints:** Balancing her freelance work with building her product meant she had to manage her time carefully.
2. **Limited Resources:** She couldn't afford expensive tools, and every dollar spent needed to count.
3. **Unfamiliar Territory:** Product development, manufacturing, and marketing were new challenges for her.

To overcome these obstacles, Sarah applied fundamental project management principles to structure her approach.

Step 1: Define the Project Objectives

Sarah's first step was to clearly outline what she wanted to achieve:

Goal: Launch her eco-friendly planner by the start of the new year.

Success Metrics:

- Produce 500 planners in time for the holiday shopping season.
- Generate $10,000 in sales during the first three months.
- Build an email list of at least 1,000 potential customers.

Step 2: Create a Simple Project Plan

Using a free project management tool, Sarah broke her project into manageable phases:

1. **Design Phase:**
   - Research competitors and gather feedback on planner features.
   - Create the planner layout and cover design.
2. **Production Phase:**
   - Source eco-friendly materials and find a reliable manufacturer.
   - Order a small batch of planners for quality testing.
3. **Marketing Phase:**
   - Build a basic website with an online store.
   - Create social media profiles and start sharing her story.
4. **Launch Phase:**
   - Host a virtual launch event to showcase the product.
   - Run targeted ads on Instagram and Facebook to drive traffic.

Step 3: Prioritize Tasks and Set Deadlines

With her phases mapped out, Sarah prioritized tasks based on impact and urgency.

Examples:

- High Priority: Finalizing the design and sourcing materials (critical path).
- Medium Priority: Building her social media presence.
- Low Priority: Refining packaging (nice-to-have but not essential).

She used time blocking to dedicate specific hours each week to her project, ensuring steady progress without overwhelming her schedule.

Sarah's Weekly Time Blocks:

Day	Time	Focus Area
Monday	7-9 PM	Design work
Wednesday	7-9 PM	Production planning
Thursday	7-9 PM	Marketing and social media
Saturday	9 AM-1 PM	Website development and admin
Sunday	2-4 PM	Planning and review

Step 4: Manage Risks and Adapt

Sarah anticipated potential risks and prepared mitigation plans:

- **Risk:** Delays in manufacturing.
- **Solution:** Build buffer time into her schedule and order samples early.
- **Risk:** Difficulty reaching her target audience.
- **Solution:** Partner with influencers in the sustainable living niche to expand her reach.

When her first manufacturer failed to meet her quality standards, Sarah quickly pivoted to a new supplier she'd researched as a backup.

Step 5: Measure Progress and Celebrate Milestones

Sarah tracked her progress weekly, checking off completed tasks and adjusting her plan as needed. She also celebrated small wins - like completing her design or gaining her first 100 email subscribers - to stay motivated.

The Results

By the time the holiday season arrived, Sarah had achieved her goals:

- Produced 500 planners on schedule.
- Sold out her first batch within two months, generating $12,000 in revenue.
- Built a loyal following of over 1,500 email subscribers and 3,000 social media followers.

Her success wasn't just about reaching her metrics - it was about proving to herself that she could turn her idea into reality with determination, creativity, and structured project management.

Lessons from Sarah's Success

1. **Clarity is Key:** Defining clear goals gave Sarah direction and focus.
2. **Break It Down:** A detailed project plan turned a daunting task into manageable steps.
3. **Prepare for Challenges:** Anticipating risks and adapting quickly kept her project on track.
4. **Celebrate Small Wins:** Acknowledging progress helped Sarah stay motivated.
5. **Start Simple:** By focusing on essential tasks and affordable tools, Sarah maximized her limited resources.

Tools Sarah Used

Tool	Purpose	Cost	URL
Trello	Task tracking and project phases	Free	trello.com
Google Sheets	Budget tracking and sales projections	Free	sheets.google.com
Canva	Design work and social media graphics	Free tier	canva.com
Shopify	E-commerce website	$29/ month	shopify.com
Mailchimp	Email marketing	Free tier	mailchimp.com

The Takeaway

Sarah's story illustrates that even solopreneurs with minimal resources can achieve big goals by applying basic project management principles. Whether you're launching a product, growing a service, or exploring a new idea, a structured approach helps you navigate challenges and make meaningful progress.

In the next chapter, we'll look at how to lead from the front - developing a leadership mindset that inspires others and strengthens every project you take on. Let's keep going.

11

The Leadership Mindset for Project Success

1. Communicating Like a Leader

Building Trust and Clarity with Your Team and Stakeholders

In startups, where timelines are tight, resources are limited, and pressure is constant, how you communicate can make or break your project. Leadership in this environment isn't just about making decisions or setting direction - it's about creating connection, clarity, and confidence through communication.

Communicating like a leader isn't about sounding impressive or being the loudest voice in the room. It's about being clear, authentic, and intentional. It's about building trust, aligning people around a shared goal, and making others feel heard, valued, and inspired.

Why Communication Is a Leadership Superpower

Great communication doesn't just transmit information - it builds relationships and drives action. Here's what effective leadership communication enables:

- **Builds Trust**: Transparency and consistency reassure your team that you're dependable and honest.
- **Drives Alignment**: Clear messaging ensures everyone is pulling in the same direction.
- **Fosters Engagement**: Open communication makes people feel valued, increasing motivation and ownership.
- **Reduces Friction**: Proactive communication prevents confusion, delays, and avoidable conflict.

The Principles of Leadership Communication

1. Be Clear and Concise

Confusion leads to missteps. As a leader, your job is to make sure your message is understood - not just heard.

How to Communicate Clearly:

- Use simple, direct language.
- Avoid jargon unless necessary - and explain it if you use it.
- Stick to one key message per communication.

Example:
Instead of saying, "Let's optimize the omnichannel workflow," say, "Let's make sure customers get a consistent experience across all platforms."

2. Listen Like a Leader

Communication isn't just about talking - it's about listening. When you genuinely listen, you show your team that their voice matters.

Tips for Active Listening:

- Make eye contact and avoid distractions.
- Summarize what you've heard to confirm understanding.
- Ask open-ended questions like, "What's your take on this?" or "What would make this easier?"

Pro Tip:
Take notes during conversations - not just to remember details, but to show you care about what's being said.

3. Communicate Frequently and Transparently

Silence creates uncertainty. Regular, transparent updates help maintain momentum and trust - especially during uncertainty or change.

How to Stay Connected:

- Hold brief weekly check-ins.
- Share progress updates, even when progress is slow.
- Be open about challenges - your team can help if they understand what's happening.

Example:
"We've hit a delay in manufacturing, but we're working on a solution. I'll share options in tomorrow's meeting and get your input."

4. Tailor Your Message to Your Audience

Your message may be the same, but how you deliver it should vary depending on who's listening.

Tailoring Tips:

- **Team members** want clarity on responsibilities and next steps.
- **Investors or executives** want high-level strategy, results, and risks.
- **Customers** want to know what's in it for them.

Example:
When talking to your developer, explain the technical scope. When updating your investor, focus on market traction and growth potential.

5. Be Authentic and Human

You don't have to have all the answers. In fact, pretending you do can hurt your credibility. Authenticity builds trust and creates stronger connections.

Ways to Be Authentic:

- Admit what you don't know - and be willing to find out.
- Speak with empathy - acknowledge emotions, not just facts.
- Share your passion and personal stake in the project.

Communicate Solutions, Not Just Problems

Even the most vigilant horizon scanners can't prevent every crisis. What separates exceptional project managers from average ones isn't whether they encounter problems - it's how quickly they pivot from problem identification to problem solving.

When issues arise, resist the urge to focus on assigning blame or simply reporting the delay. Instead, immediately shift into solution mode: gather the right people, define the problem clearly, identify multiple potential approaches, and implement the most promising one with decisive action. Develop a reputation as someone who brings solutions, not just problems, to stakeholders.

This problem-solving orientation does more than save individual projects - it builds the resilience and adaptability that emerging businesses need to survive and thrive in uncertain environments.

Fostering a Culture of Open Communication

Great leaders don't just communicate well - they create environments where communication thrives.

How to Build a Communicative Culture:

- Encourage questions and feedback - even the tough ones.
- Normalize asking for help and sharing challenges.
- Act on feedback to show it matters.

Example:
Start team meetings with, "What's something we should improve this week?" or "What's holding us back right now?"

Communication Tools for Project Leaders

The right tools can enhance your communication effectiveness. Here are some options that work well for startups:

Tool	Best For	Features	URL
Slack	Day-to-day team communication	Channels, direct messaging, integrations	slack.com
Zoom	Video meetings and presentations	Screen sharing, recording, breakout rooms	zoom.us
Loom	Asynchronous video updates	Screen recording, easy sharing	loom.com
Notion	Documentation and knowledge sharing	Collaborative editing, templates	notion.so
Miro	Visual collaboration	Virtual whiteboards, templates	miro.com

Overcoming Common Communication Pitfalls

- **Mixed Messages**: Ensure consistency between your words, tone, and actions.
- **Overloading Your Team**: Keep messages focused. Don't bury the point in too much detail.
- **Avoiding Difficult Conversations**: Tackle issues directly and constructively - early and often.

Real-World Example: Sarah's Planner Launch (from Chapter 10)

When Sarah launched her eco-friendly planner business, her success wasn't just about a great product - it was about how she communicated:

- She **kept her manufacturer aligned** with clear expectations and follow-ups.
- She **connected with her audience** by sharing her story and progress on social media.

- She **stayed self-aware**, tracking her goals and using feedback to guide her next steps.

Outcome: Her ability to communicate effectively helped her navigate delays, engage her audience, and deliver a successful launch. Her calm, consistent, and transparent communication helped her overcome setbacks and build strong relationships.

The Takeaway

Communicating like a leader means more than giving instructions - it means connecting with people. It means being clear, authentic, and intentional when things are confusing, honest when things are tough, and encouraging when energy is low. By tailoring your messages, listening actively, and fostering open dialogue, you can build trust, alignment, and engagement within your team and among stakeholders.

Leadership isn't just about what you say - it's about how you make people feel and how effectively you connect with them. Whether you're managing a project, leading a team, or launching your startup solo, leadership communication is your most powerful tool.

In the next section, we'll explore how to lead through conflict and collaboration - so you can keep your team working together, even when tensions rise. Let's keep building your leadership mindset!

2. Fostering Collaboration

Techniques to Encourage Teamwork and Ownership of Tasks

Leadership is rarely a solo act. In the fast-moving world of startups and project-driven teams, collaboration is what transforms a handful of capable individuals into a cohesive, high-performing unit. True collaboration is more than just working together - it's a culture, a mindset, and a daily habit that leaders must intentionally nurture.

In this section, we'll look at practical ways to foster real collaboration in your team. From setting a shared vision to creating an environment of mutual respect and ownership, these strategies can help your team move in sync, support one another, and deliver better results - together.

Why Collaboration Matters

Startups thrive on speed, creativity, and adaptability - all of which depend on how well your team collaborates. When collaboration is strong:

- **Problems are solved faster** because people bring different perspectives.
- **Workflows are smoother** thanks to shared ownership and mutual understanding.
- **Team morale is higher**, as people feel heard, valued, and supported.
- **Results improve**, as more people contribute meaningfully to a shared vision.

1. Create a Shared Vision

Before a team can move forward together, they need to know where they're going - and why it matters.

How to Set a Shared Vision:

- **Define the "Why"**: What are you trying to achieve? Why is it important to your users, company, or community?
- **Make it Collaborative**: Involve the team in shaping goals to increase buy-in.
- **Reinforce Often**: Revisit the vision during meetings and milestones to keep it front of mind.

Example:
"This feature isn't just a cool add-on - it's the key to cutting customer churn by 20%, which helps us grow sustainably."

2. Build Trust and Psychological Safety

Trust is the foundation of collaboration. When people feel safe to speak up, share ideas, or admit mistakes, innovation and accountability flourish.

How to Build Trust:

- Be transparent about challenges and decisions.
- Admit your own mistakes - this encourages others to do the same.
- Encourage respectful disagreement, and model how to listen without judgment.

Pro Tip:
Start meetings by asking, "What's something that's going well? What's something we could improve together?"

3. Define Roles and Responsibilities Clearly

Ambiguity kills momentum. Clear roles empower individuals to contribute confidently and avoid stepping on toes.

Tools You Can Use:

- **RACI Matrix**: Clarifies who is *Responsible*, *Accountable*, *Consulted*, and *Informed* for each task.
- **Task Ownership in Project Tools**: Assign roles directly in Trello, Asana, or Notion to maintain visibility.

Watch For:
Overlap or gaps in responsibility - these often lead to miscommunication or duplicated effort.

4. Encourage Open and Ongoing Communication

Collaboration requires continuous dialogue - not just occasional status updates.

Ways to Encourage Communication:

- **Daily or Weekly Stand-Ups**: Short, structured check-ins to align on priorities and roadblocks.
- **Collaboration Tools**: Use Slack, Teams, or Notion to keep conversations and documents accessible.
- **Anonymous Feedback Options**: For teams that are still building trust, anonymous feedback tools can surface issues early.

Pro Tip:
Intentionally include quieter team members in conversations. Invite their thoughts directly or in smaller breakout sessions.

5. Promote Ownership and Autonomy

People are more engaged when they feel their work matters. That sense of ownership drives better outcomes and stronger commitment.

How to Foster Ownership:

- Let people own decisions in their area of expertise.
- Connect individual tasks to larger outcomes.
- Avoid micromanaging - support your team without hovering.

Example:
Instead of assigning "Write the report," say, "Your analysis will be key to securing funding - how do you want to approach it?"

6. Solve Problems as a Team

Setbacks are inevitable - but when your team rallies around a problem instead of finger-pointing, you build resilience and trust.

Collaborative Problem-Solving Tips:

- Use group brainstorming or retrospectives to uncover solutions.
- Invite cross-functional input for richer ideas.
- Shift the mindset from blame to "What can we learn?"

Real-World Example: Building a Collaborative Team

When launching her eco-friendly planner (Chapter 10), Sarah worked with freelancers and vendors across time zones and skill sets.

Here's how she made collaboration work:

- **She defined clear goals** for the design, production, and marketing teams.
- **She held regular check-ins**, despite working asynchronously.
- **She focused on strengths**, allowing her designer to lead creative and her manufacturer to handle materials and logistics.

The Result:
Despite limited resources, the team delivered on time and created a product that resonated with customers - because everyone knew their role, felt heard, and stayed connected.

Common Collaboration Pitfalls and Fixes

Pitfall	How to Avoid It
Lack of clarity	Use clear goals and visible task ownership.
One voice dominates	Facilitate meetings so everyone contributes.
Team silos	Encourage cross-functional working groups.
Avoiding conflict	Address issues early with curiosity and solutions in mind.

The Takeaway

Collaboration is more than cooperation - it's about building a culture where people feel safe, supported, and motivated to contribute their best. As a leader, your role is to create the conditions for that to happen by clarifying goals, encouraging communication, and reinforcing shared success.

In the next section, we'll tackle conflict head-on - because even great teams will disagree. And that's not only okay - it can be a source of strength. Let's explore how to manage conflict constructively and keep your team moving forward.

3. Developing Resilience

Staying Motivated and Inspiring Others During Tough Times

Resilience is the cornerstone of effective leadership, especially in emerging businesses' high-pressure, high-uncertainty world. It's what keeps you moving forward when setbacks hit, what inspires your team when morale dips, and what sustains your vision when the road ahead feels uncertain.

Developing resilience isn't about pretending everything is fine or ignoring challenges. It's about embracing reality, staying adaptable, and fostering a mindset that turns adversity into opportunity. In this section, we'll explore what resilience looks like in startup leadership, how to develop it in yourself, and how to foster it in your team - so that when the going gets tough (and it will), you can keep going strong.

Why Resilience Is a Leadership Essential

Resilient leaders don't avoid adversity - they respond to it constructively. They're able to:

- **Stay calm and make smart decisions**, even under pressure.
- **Inspire their teams**, especially during uncertain or difficult moments.
- **Bounce back from setbacks**, turning failure into growth.
- **Keep moving forward**, staying connected to their vision without losing sight of the present.

1. Cultivate a Resilient Mindset

Resilience starts with mindset. Leaders who bounce back from challenges tend to:

- See obstacles as opportunities to grow.
- Focus on progress over perfection.
- Accept that setbacks are part of the journey.

Practical Ways to Build a Resilient Mindset:

- **Reframe setbacks**: Instead of "We failed," try "We learned something we can use."
- **Focus on your circle of control**: Spend your energy on what you can influence, not what you can't.
- **Practice gratitude**: Reflect regularly on what's working, even during tough times.

2. Fuel Motivation with Small Wins

Long-term goals can feel distant - especially during slow or difficult periods. That's where small wins come in.

Why It Works: Small wins create a sense of momentum and reinforce a feeling of competence. They remind you (and your team) that progress is being made - even if it's one step at a time.

How to Use Small Wins:

- Break big goals into short-term milestones.
- Track progress visually (e.g., a Kanban board, checklist, or shared dashboard).
- Celebrate small victories publicly - shout-outs, emojis, coffee runs, or a simple "Well done!"

Example:
Instead of focusing only on the final product launch, celebrate reaching prototype approval, passing user testing, or uploading the final design assets.

3. Build a Resilient Team Culture

You set the tone. If you handle challenges with calm and confidence, your team is more likely to do the same.

Ways to Foster Team Resilience:

- **Normalize learning from failure**: Talk openly about what went wrong and what was learned.
- **Support each other**: Encourage empathy, check-ins, and team debriefs after challenges.
- **Involve the team in problem-solving**: When people help create the solution, they're more committed to seeing it through.

Pro Tip:
After a tough sprint, hold a retrospective not just on tasks - but on how the team handled pressure and what would make it easier next time.

4. Develop Personal Resilience Habits

Resilience doesn't mean grinding non-stop. It's about recovery as much as effort.

Self-Care for Founders and Leaders:

- **Sleep**: You can't think clearly or lead well when you're exhausted.
- **Exercise**: Even 15 minutes a day helps reduce stress and improve focus.

- **Mindfulness**: Journaling, deep breathing, or meditation builds mental clarity.
- **Connection**: Talk to someone you trust - a mentor, peer, friend, or advisor.

Don't go it alone.
Even the best leaders need support, perspective, and space to regroup.

5. Stay Connected to Your "Why"

Purpose fuels resilience. When the work gets hard, reminding yourself (and your team) of *why* it matters can be the boost you need to keep going.

Ways to Stay Grounded in Purpose:

- Revisit your mission or vision statement.
- Share customer stories or testimonials with the team.
- Keep a visual or physical reminder of the impact you're aiming to make.

Example:
If you're building a mental wellness app, post screenshots of user feedback thanking you for the help it's provided. That impact is your anchor.

Real-World Example: Resilience in Action

Sarah, the solopreneur from Chapter 10, faced a major challenge when her manufacturer backed out weeks before her eco-friendly planner launch.

Instead of panicking:

- She **reframed the delay** as time to polish her marketing plan.
- She **found a new supplier**, drawing on contacts she'd built as a backup.
- She **kept her customers informed**, building trust through transparency.
- She **leaned on her purpose**, reminding herself this wasn't just a planner - it was about helping people live more intentionally.

Result: Sarah still launched successfully, with an even stronger relationship with her audience and valuable lessons for the next round.

Common Pitfalls and How to Avoid Them

Pitfall	How to Overcome It
Overworking	Build recovery into your schedule - rest is a performance tool, not a reward.
Isolation	Share your struggles with trusted peers or advisors. You're not alone.
Perfectionism	Focus on progress, not perfection. Done is better than perfect in high-stress environments.
Avoiding Emotions	Name what you're feeling, then use it as fuel for constructive action.

The Takeaway

Resilience is the leadership superpower that keeps startups alive - and growing. It's not about being unshakable or fearless. It's about staying grounded, choosing action over despair, and helping others do the same.

As you lead through uncertainty, remember:

- Resilience grows with practice.
- Small wins and support systems matter.
- Your mindset sets the tone for your team.

In the next chapter, we'll explore how to scale your success without losing the clarity, culture, and connection you've worked so hard to build. Let's keep going - stronger and more resilient than ever.

12

Taking Your Skills to the Next Level

1. Building a Culture of Continuous Improvement

Turning Lessons into Lasting Success

Great project managers know that success isn't a finish line - it's a moving target. The best teams and organizations thrive not by staying still but by continually learning, adapting, and improving. This culture of continuous improvement helps your team deliver better results, stay agile, respond to change effectively and keep engaged, motivated, and competitive in fast-changing environments.

Building a culture of continuous improvement isn't about chasing perfection - it's about creating an environment where learning is valued, mistakes are embraced as opportunities, and everyone feels empowered to contribute to growth. In this section, we'll explore how to embed this culture into your startup or team, turning every project into a stepping stone toward greater success.

This section is all about building that culture - one where feedback flows freely, experiments are encouraged, and small changes add up to big results over time.

Why Continuous Improvement Matters

Startups don't have time to waste. The ability to learn quickly and improve consistently is a strategic advantage.

Benefits of a Continuous Improvement Culture:

- **More Efficiency**: Eliminate bottlenecks and optimize how your team works.
- **Faster Innovation**: Iterate faster, test smarter, and refine what works.
- **Higher Engagement**: People feel invested when their feedback leads to change.
- **Greater Resilience**: Teams that reflect and adapt bounce back faster from setbacks.

Principles of Continuous Improvement

1. Start with Reflection

Improvement begins with self-awareness. Without pausing to assess what's working and what's not, you risk repeating mistakes or missing growth opportunities.

How to Reflect as a Team:

- **Run Retrospectives**: After key milestones or sprints, ask:
  "What went well?"
  "What didn't?"
  "What should we try next time?"
- **Use Data**: Back up reflection with key metrics - velocity, engagement, NPS, delivery timelines.
- **Make it a Habit**: Schedule regular reflection time, even mid-project.

Pro Tip: Keep a running "Lessons Learned" document. Revisit it before kicking off new initiatives.

2. Foster a Growth Mindset

Continuous improvement only thrives in teams that believe improvement is possible.

How to Encourage Growth Thinking:

- **Celebrate learning, not just results**: Applaud thoughtful experimentation - even if it didn't go to plan.
- **Talk openly about failures**: Reframe them as stepping stones, not black marks.
- **Be transparent about your own growth**: Model what it means to evolve and adapt.

Example:
"We tried launching on Tuesday and saw a 20% drop in signups. That tells us something important - we'll test a new launch window next week."

Pro Tip: Lead by example - share your own learning experiences openly with your team.

3. Empower Your Team To Improve

Improvement isn't a top-down directive - it works best when everyone contributes.

How to Empower:

- **Ask for ideas**: "What part of our process slows you down?" or "What's one thing we could improve this month?"

- **Act on feedback**: Close the loop by testing or implementing suggestions.
- **Offer training or tools**: Give your team access to the resources they need to build skills and optimize how they work.

Example: "Our current project tracking system feels clunky. Does anyone have ideas for tools or processes that might work better?"

Pro Tip: Create space for regular improvement conversations - during standups, team huddles, or one-on-ones.

4. Build Feedback Loops That Actually Work

Improvements stick when they're part of an ongoing feedback cycle.

How to Create Feedback Loops:

- **Use the PDCA cycle** (Plan → Do → Check → Act) to test and refine changes.
- **Track changes** in a shared doc or board so everyone sees what's evolving.
- **Debrief often**: Ask, "Did this change help? What should we do next?"

Example:
If your sprint reviews are running long, try shortening them by 15 minutes. Ask afterwards if the change improved focus. If yes, keep it. If not, iterate again.

Pro Tip: Use the **Plan-Do-Check-Act (PDCA)** cycle to structure continuous improvement efforts:

1. **Plan:** Identify an area to improve.
2. **Do:** Implement a small change or experiment.
3. **Check:** Evaluate the results.
4. **Act:** Standardize successful changes or refine further.

Phase	Description	Questions to Ask
Plan	Identify an area to improve	What needs to change? What's our goal?
Do	Implement a small change or experiment	How can we test this on a small scale?
Check	Evaluate the results	Did it work? What did we learn?
Act	Standardize successful changes or refine further	How do we scale or adjust based on results?

5. Recognize and Reward Progress

Even small improvements deserve recognition - because they add up over time.

Ways to Celebrate Growth:

- Highlight improvements in team updates or demos.
- Give shout-outs to individuals who identify or lead meaningful changes.
- Document and share before/after stories - especially when a small tweak yields big results.

Example:
"Thanks to Maria's suggestion, we switched to async updates on Mondays. We've saved over 3 hours per month in meetings - nice work!"

Example: "Thanks to Alex's suggestion, our new onboarding process has cut setup time in half. Great work!"

Real-World Examples: Continuous Improvement in Action

Case Study 1: Improving Customer Onboarding in a SaaS Startup

A startup offering project management tools was losing trial users before they completed onboarding.

What They Did:

1. **Reflected:**
   Surveyed drop-offs to understand where users disengaged.
2. **Involved the Team:**
   The support and product teams brainstormed improvements.
3. **Tested a Fix:**
   Introduced a 2-minute guided walkthrough.
4. **Measured Results:**
   Completion rates jumped from 42% to 73%.
5. **Scaled It:**
   Rolled out the walkthrough across all entry points and built a dashboard to monitor engagement.

Outcome:
A small, targeted improvement drove higher conversion rates, better customer satisfaction, and team-wide alignment around experimentation.

Case Study 2: A Startup Streamlines Its Customer Support Process

A SaaS startup struggled with slow response times to customer inquiries, leading to declining satisfaction rates.

What They Did:

1. **Reflected:**
   Conducted a retrospective and identified bottlenecks in their support workflow.
2. **Involved the Team:**
   Asked support agents for ideas to improve response times.
3. **Experimented:**
   Tested an AI-powered ticketing system to categorize and prioritize inquiries automatically.
4. **Checked Results:**
   Measured the impact, finding a 40% reduction in response times and increased customer satisfaction.
5. **Scaled and Standardized:**
   Rolled out the new system across the entire team and continued to monitor its performance.

Outcome:
By involving the team and embracing small, iterative changes, the startup transformed a challenge into an opportunity to improve its customer experience.

Common Pitfalls - and How to Avoid Them

Pitfall	Solution
Resistance to change	Involve the team early and explain the "why" behind improvements.
Trying to fix everything at once	Focus on one meaningful improvement at a time.
Ignoring feedback	Close the loop - show that feedback leads to action.
Treating improvement as "extra" work	Make it part of how work gets done - not a separate initiative.

The Takeaway

Continuous improvement is a habit, not a one-time effort. It starts with reflection, thrives on feedback, and grows when every team member feels empowered to contribute.

The best leaders don't just deliver results - they build teams that keep getting better. When learning becomes part of your culture, progress never stops.

In the next section, we'll look ahead to the future of project management - exploring the tools, trends, and strategies shaping the next era of high-performance ventures.

2. The Future of Project Management

Trends and Tools Shaping the Next Generation of Leaders

Project management is evolving - fast. New technologies, shifting team structures, and changing expectations are redefining how projects are led and delivered. For founders, project leaders, and team managers, staying ahead of these changes isn't just helpful - it's essential for maintaining a competitive edge.

This section explores the key trends transforming project management today and tomorrow. By understanding these shifts and embracing new tools, you'll future-proof your skills and position yourself as a forward-thinking leader, ready for what comes next.

Key Trends Shaping the Future

1. Artificial Intelligence (AI) and Automation Are Reshaping the Landscape

Artificial intelligence and automation are no longer futuristic concepts - they're now everyday tools. These technologies are reducing manual effort, streamlining operations, and uncovering insights that help leaders make smarter decisions.

Where AI Is Making an Impact:

- **Task Automation**: Tools like Asana, ClickUp, and Monday.com use AI to prioritize work, set deadlines, and manage dependencies.
- **Predictive Analytics**: AI-driven platforms analyze past performance to forecast risks, identify bottlenecks, and suggest proactive adjustments.
- **Automated Reporting**: Dashboards generate real-time insights without manual effort.

How to Get Started:

- Integrate AI tools to automate scheduling, reporting, or task tracking.
- Use machine learning features in your project software to spot trends or flag risks early.
- Train your team on how to collaborate with AI - not just use it.

Pro Tip: Start small. Automate one repetitive process by integrating AI tools into specific aspects of your workflow, such as scheduling or reporting, and track the time or error reduction it brings.

Tools Available Now

AI Tool	What It Does	Startup Benefit	URL
Asana's Workload	Balances team assignments	Prevents burnout and optimizes resources	asana.com/features/workload
ClickUp AI	Writes summaries, generates tasks	Speeds up documentation and planning	clickup.com/ai
Forecast	Predictive project planning	More accurate estimations	forecast.app
Lili.ai	Risk identification and management	Early problem detection	lili.ai

2. Remote, Hybrid, and Flexible Teams Are the New Normal

The traditional 9-to-5, co-located team is becoming increasingly rare. Startups and global teams now operate across time zones and platforms, requiring new ways to lead and connect.

What's Changing:

- **Asynchronous Communication**: Teams use tools like Loom, Slack, and Notion to collaborate without being online at the same time.
- **Decentralized Teams**: Cross-functional, often globally distributed teams now work together fluidly.
- **Gig and Freelance Talent**: Many projects now rely on specialists who work contractually or part-time.

How to Lead Effectively:

- Emphasize results, not hours worked.
- Use clear documentation and task boards (e.g. Trello, Basecamp) to ensure alignment.
- Foster team belonging through regular check-ins and inclusive practices.

Example: A project manager might use Trello to assign tasks and run a remote stand-up using Loom video updates while tracking all deliverables on Notion. This allows team members in different time zones to contribute effectively and stay connected while working flexibly.

3. Agile, Hybrid, and Evolving Methodologies

Agile has reshaped how teams build products, but it's not one-size-fits-all. Increasingly, leaders are blending Agile with traditional models to meet unique needs.

Key Trends:

- **Scaled Agile (SAFe, LeSS)**: Enables larger teams or organizations to maintain agility while coordinating across multiple teams.

- **Hybrid Project Management**: Combines Agile principles (like sprints) with Waterfall-style milestones or budgets.
- **Continuous Feedback Loops**: Reflection and adjustment are built into every phase - not just post-launch.

How to Adapt:

- Learn the strengths and weaknesses of Agile, Scrum, Kanban, and Waterfall.
- Tailor your methodology to fit your team's size, pace, and project complexity.
- Revisit your process after each project or sprint - improvement is iterative.

4. The Human Side Is More Important Than Ever

While technology evolves rapidly, human leadership remains irreplaceable. Emotional intelligence (EQ), empathy, and communication are now the defining traits of effective project leaders.

Why Emotional Intelligence Matters:

- **Boosts Trust**: Teams follow leaders who listen, understand, and support them.
- **Improves Performance**: Teams feel safer taking risks and speaking up.
- **Reduces Burnout**: A leader who balances ambition with compassion builds healthier, more sustainable teams.

What This Looks Like in Practice:

- Checking in personally with remote team members.
- Recognizing burnout early and adjusting workloads.
- Celebrating progress as well as results.

Example: A leader might notice increased stress on the team mid-sprint and organize a "no-meeting" afternoon or a casual check-in to reset and reconnect. They adjust workloads as needed to support their team's well-being.

5. Data-Driven Decision-Making

The most effective project leaders use real-time data - not gut instinct - to drive progress, mitigate risk, and adapt quickly.

The Role of Data Today:

- **KPI Dashboards**: Metrics like velocity, burn rate, and completion rates guide decision-making.

- **Budget and Timeline Forecasting**: AI tools can now predict slippage or budget overruns before they occur.
- **Qualitative + Quantitative**: Smart leaders blend data insights with team feedback for a fuller picture.

How to Stay Data-Ready:

- Use tools like Jira, Power BI, or Tableau to visualize your metrics.
- Establish clear success indicators before launching any project.
- Review metrics weekly and adjust plans based on what you see.

Pro Tip: Avoid "vanity metrics." Focus on data that directly informs decision-making (e.g., conversion rates, retention, delivery cycle time) and adjust your strategy to stay proactive rather than reactive.

The Evolving Role of the Project Manager

The project manager of the future isn't just a task organizer - they're a strategic leader who:

- **Fosters Collaboration**: Facilitates team synergy in diverse, distributed environments.
- **Embraces Innovation**: Adopts new tools and practices to enhance efficiency.
- **Balances Technology and Humanity**: Leverages data and AI while maintaining a people-first approach.

Example: A project manager might use predictive analytics to identify potential delays while using emotional intelligence to navigate team dynamics and maintain morale.

The Evolution of Project Management Roles

As technology advances, project management roles are diversifying:

Traditional Role	Emerging Evolution	Key Skills Needed
Project Manager	Digital Project Manager	Tech fluency, remote team leadership
Coordinator	Automation Specialist	AI/ML understanding, process optimization
Scheduler	Resource Strategist	Predictive analytics, capacity planning
Reporter	Data Visualization Expert	Dashboard creation, data storytelling
Task Assigner	Team Coach/Facilitator	Emotional intelligence, servant leadership

Preparing for the Future

To stay ahead in this evolving landscape, focus on continuous learning and adaptability:

1. Upskill Regularly:

- Take courses in emerging technologies like AI, data analytics, and blockchain.
- Resources:
    - Coursera - Offers PM certification and tech courses
    - LinkedIn Learning - Practical PM skills
    - Product School - Product management focus

2. Develop Soft Skills:

- Enhance your emotional intelligence, communication, and leadership abilities.
- Resources:
    - MindTools - Leadership development
    - Toastmasters - Communication skills

3. Stay Curious:

- Follow industry trends through blogs, webinars, and professional networks.
- Resources:
    - ProjectManagement.com - PMI resources
    - Digital Project Manager - Modern PM trends

4. Experiment with New Tools:

- Adopt a test-and-learn approach to integrate innovative solutions into your projects.
- Try one new tool per quarter to gradually expand your toolkit.

The Changing Role of the Project Manager

Today's project leaders wear many hats. They're part strategist, part coach, and part technologist.

Tomorrow's Project Manager Will:

- Facilitate team synergy across tools, time zones, and personalities.
- Drive innovation while staying grounded in human connection.
- Blend EQ with IQ - balancing data, tech, and people leadership.

Final Thought: The tools will change, but the essence of leadership - clarity, empathy, adaptability - will remain timeless.

The Takeaway

Project management is entering a new era - one defined by smart tools, hybrid teams, and human-centered leadership. By embracing these trends and building your skills around both technology and people, you'll be equipped not just to manage projects - but to lead change.

In the next section, we'll wrap up this journey with practical ways to scale your project management skills and apply them to bigger, more complex challenges.

3. Next Steps for Readers

Turning Knowledge into Action

You've learned the principles, tools, and techniques that form the foundation of effective project management for startups and emerging businesses. Now comes the most important step: putting this knowledge into practice. Project management isn't just a skill - it's a discipline that gets stronger with every project you tackle. By taking deliberate, actionable steps, you can continue to refine your abilities and lead your team to greater success.

This section outlines your next steps for applying what you've learned, growing your skills, and creating a lasting impact in your projects and beyond.

Step 1: Start Small, Start Now

The best way to build momentum is to begin with a manageable project.

What to Do:

- Choose a current or upcoming project to apply your new skills.
- Use simple tools, such as a task board or checklist, to track progress.
- Focus on mastering one or two techniques at a time, like prioritization or stakeholder alignment.

Example: If you're planning a product launch, create a project charter that defines goals, timelines, and responsibilities.

Step 2: Experiment and Reflect

Project management is not one-size-fits-all. Experiment with different methods, tools, and frameworks to find what works best for you and your team.

How to Experiment:

- Try a new tool like Asana or Notion for task tracking.
- Incorporate a framework like Agile or Kanban into your workflow.
- Adjust meeting formats or reporting styles to suit your team's preferences.

Reflection Tips:

- After each project, conduct a retrospective to identify what worked and what didn't.
- Ask for feedback from your team or stakeholders to gain fresh perspectives.

- Document lessons learned and apply them to future projects.

Step 3: Focus on Personal Development

Great project managers are always learning. Dedicate time to growing both your technical and leadership skills.

What to Learn:

- **Technical Skills:** Familiarize yourself with new tools, methodologies, or industry-specific knowledge.
- **Leadership Skills:** Improve communication, emotional intelligence, and decision-making abilities.

Resources to Explore:

- Online courses (e.g., Coursera, Udemy, LinkedIn Learning).
- Books on leadership, project management, or innovation.
- Professional certifications, such as PMP (Project Management Professional) or Agile Certified Practitioner.

Pro Tip: Set a personal development goal, like mastering a new tool or completing a certification, within the next six months.

Step 4: Build a Support Network

Surround yourself with mentors, peers, and communities that can help you grow.

How to Build Your Network:

- Join startup and project management groups on platforms like LinkedIn or Meetup.
- Attend industry events, workshops, or webinars to connect with like-minded professionals.
- Seek mentorship from experienced leaders who can offer guidance and insights.

Example: A mentor might help you navigate complex challenges, while a peer group can share ideas and best practices.

Step 5: Embrace a Growth Mindset

Every project - whether a triumph or a trial - is an opportunity to learn. By embracing a growth mindset, you'll approach challenges with curiosity and determination.

How to Cultivate a Growth Mindset:

- Reframe setbacks as opportunities to improve.
- Celebrate progress, not just outcomes.
- Stay open to feedback and new ideas.

Pro Tip: Keep a journal or log to document your learnings from each project. Reviewing these entries can inspire confidence and show how far you've come.

Step 6: Share Your Knowledge

Teaching others solidifies your own understanding and expands your impact.

Ways to Share:

- Mentor a colleague or team member interested in project management.
- Write articles or blog posts sharing your experiences and insights.
- Offer to lead workshops or training sessions for your team.

Example: Share how you streamlined a process or overcame a challenge during a team meeting or on LinkedIn.

Your Challenge: Take the First Step

Start today by identifying one immediate action you can take to improve how you manage projects. Maybe it's revising your next team meeting agenda to focus on priorities, trying out a new tool, or defining clearer goals for your current project.

Remember: Progress is built on small, consistent actions.

The Takeaway

Project management is a journey, not a destination. By starting small, reflecting on your experiences, and committing to ongoing growth, you'll develop the skills and confidence to lead projects - and people - with excellence. The tools and techniques you've learned here are just the beginning. Now it's time to take them into the real world, make them your own, and create your own success stories.

Your next project could be the one that changes everything - so let's get started!

Conclusion ::
Your First Project Management Success Story

---

1. Take the First Step and Start Your Journey

Every Great Success Begins with a Single Step

You've reached the end of this book, but your journey is just beginning.

Mastering project management isn't about memorizing every tool or following a rigid formula - it's about learning by doing, staying adaptable, and taking purposeful action. Whether your next project is launching a new product, organizing a team effort, or simply streamlining and managing your daily workflow, the most important thing is this: **start**.

Every successful leader you've read about - every founder, team lead, project manager, or solopreneur - began where you are now: with a goal, a bit of uncertainty, and a willingness to learn. Their journeys began with one thing in common - **they decided to act**.

Now it's your turn.

The Power of Taking the First Step

Starting something new can feel intimidating, especially when stakes are high or uncertainty is present. But the secret to success isn't perfection - it's **progress**.

When you take action, even in small doses, momentum builds. Confidence grows. Clarity sharpens. One project leads to another, and before you know it, you've built a track record of success.

Why Taking the First Step Matters:

- **Momentum:** Even a small win builds energy and excitement for the next step.
- **Clarity:** Action reveals what matters most - and what doesn't.
- **Growth:** You learn more by doing than by waiting for the perfect plan.

How to Take the First Step

1. Choose a Project That Inspires You

Pick something you care about - a challenge or opportunity that feels meaningful to you ir your team.

Examples:

- Launch a new product or feature.
- Plan an event or customer engagement initiative.
- Improve an internal process that frustrates your team.

2. Apply What You've Learned

Start by using the principles, tools, and techniques covered in this book.

Steps to Get Started:

- Define clear objectives for your project.
- Use a simple tool, like a task board or project charter, to organize your plan.
- Schedule your first check-in or meeting to align your team and set expectations.

3. Embrace Progress Over Perfection

Don't wait for all the stars to align. Start with what you have and refine along the way.
Pro Tip: Begin with small, manageable tasks to build confidence and gain early momentum.

From First Step to Success: A Simple Pathway

Starting Point	Initial Action	Near-Term Goal	Long-Term Vision
An Idea	Create a one-page project charter	Complete a proof of concept	Launch a viable product
A Problem	Map the current process	Implement one improvement	Transform the entire workflow
A Team	Hold a kickoff meeting	Complete first milestone	Build a high-performing unit
A Skill Gap	Learn one new technique	Apply it to a current project	Become a valued PM leader

The Journey Ahead

As you step into your first project, remember that challenges and setbacks are part of the process. They're not signs of failure - they're opportunities to learn and grow. Each project you complete will sharpen your skills, expand your knowledge, and strengthen your ability to lead with confidence.

The road ahead will have its twists and turns, but every step you take brings you closer to your goals. More importantly, it builds your story - your unique journey as a project manager and leader.

The Takeaway

Starting is the hardest part - but it's also the most powerful.

With the tools, insights, and strategies you now have, you're more than ready. Take that first step. Start your first project. Put your skills into action.

Remember, every great success story begins with a single moment of action. Make today that moment.

Your first project management success story is waiting. Let's write it together. Now, go build something amazing.

2. Call to Action: Apply One Tool or Template Immediately

Turn Knowledge into Action - Today

Reading about project management is valuable, but its true power lies in applying what you've learned.

You've explored the strategies, tools, and mindsets that fuel successful projects. Now it's time to put them into practice. Knowledge is powerful, but action is transformational. Your journey starts not with a grand gesture - but with one small, focused step.

Let this be the moment you move from learning to doing.

Why Start Now?

The sooner you act, the sooner your project management skills become real. Taking that first action unlocks momentum, builds confidence, and starts your personal success story.

What You'll Gain by Acting Today:

- **Momentum:** One small step leads to another - quickly building progress.
- **Confidence:** Real-world experience validates your abilities.
- **Clarity:** Putting tools into practice reveals what works for you.
- **Results:** You don't need to wait for success - you can create it now.

Step 1: Choose One Tool or Template

Review the techniques from this book and choose just one that fits your current needs. Here are a few ideas:

- **Project Charter Template**
  Define the scope, objectives, timeline, and key stakeholders of your next initiative.
- **Task Board (Kanban)**
  Set up a simple task board in Trello, Notion, or on paper to visualize progress.
- **Retrospective Template**
  Run a quick project review to capture what worked and what didn't.
- **Risk Assessment Matrix**
  Identify potential risks and plan proactive mitigation strategies.
- **Stakeholder Map**
  Clarify who's involved, their influence, and how to keep them engaged.

Examples:

- Use a project charter to define your objectives, stakeholders, and scope.
- Create a Kanban board to visualize and track tasks.
- Apply the Eisenhower Matrix to prioritize tasks by urgency and importance.

If You're Struggling With...	Try This Tool or Template
Unclear project direction	Project Charter (Chapter 2)
Task overwhelm	Eisenhower Matrix (Chapter 7)
Stakeholder confusion	RACI Matrix (Chapter 2)
Meeting inefficiency	Meeting Agenda Template (Chapter 9)
Progress tracking	Simple Kanban Board (Chapter 4)
Risk management	Risk Register (Chapter 9)

Step 2: Apply It to a Real Project

Pick a real, manageable project - big or small - and use your chosen tool to bring it to life.

How to Apply It:

- Identify the project (e.g., launching a blog, onboarding a new client, organizing a team meeting).
- Use your tool to add structure and clarity.
- Share your plan with collaborators or simply walk through it yourself.

Example:
You're planning a team offsite.

- Use the **Project Charter Template** to outline the purpose, agenda, and goals.
- Share it with your team to align expectations.
- Track tasks with a simple Kanban board.

Step 3: Focus on Progress, Not Perfection

After implementation, take 15 minutes to reflect:

- What worked well?
- What could be improved?
- How did the tool help?

Pro Tip: Don't worry if it's not perfect. The goal is progress, not perfection. Iterate and refine as you go.

Start Building Your Story

Imagine this moment a month from now:

- You've successfully used a project management tool.
- Your project ran smoother.
- You gained clarity, confidence, and real results.

That's your **first success story** - one you created with a simple decision to act.

The Takeaway

Your challenge is simple:
Choose one tool. Apply it to one project. Start today.

It could be a side hustle, a workplace initiative, or a personal goal. No project is too small to benefit from structure and clarity.

You've got the knowledge. You've got the tools.
Now it's time to take action - and write your own success story.

Let's build something extraordinary together.

Additional Features

1. Quick Start Guide

A One-Page Checklist to Jumpstart Your Project

Getting started with project management doesn't have to be overwhelming. This quick-start guide distills the essential steps and tools into a simple, actionable checklist. Whether you're managing a small team, tackling a solo project, or launching a full-scale startup initiative, this guide will help you hit the ground running.

Step 1: Define Your Project

- **Clarify Objectives:**
    - What are you trying to achieve? Define your goals clearly.
- **Identify Deliverables:**
    - What are the tangible outputs of this project?
- **Set Success Metrics:**
    - How will you measure success (e.g., revenue, user feedback, timeline)?

Step 2: Organize Your Team and Stakeholders

- **Identify Roles:**
    - Who is responsible for what? Assign roles and responsibilities.
- **Align Stakeholders:**
    - Schedule a meeting to align expectations and share the project's purpose.
- **Establish Communication Channels:**
    - Choose tools like Slack, email, or team meetings for updates.

Step 3: Plan the Project

- **Create a Timeline:**
  - Break the project into phases or sprints with clear deadlines.
- **Prioritize Tasks:**
  - Use a tool like the Eisenhower Matrix or Kanban to sort tasks by urgency and importance.
- **Choose Tools and Templates:**
  - Use a project charter, Gantt chart, or Kanban board to track progress.

Step 4: Execute with Focus

- **Hold a Kickoff Meeting:**
  - Ensure everyone understands their tasks, deadlines, and dependencies.
- **Track Progress:**
  - Use tools like Trello, Asana, or Notion to monitor tasks and milestones.
- **Encourage Collaboration:**
  - Foster teamwork through regular check-ins and open communication.

Step 5: Monitor and Adjust

- **Review Progress Regularly:**
  - Schedule weekly check-ins to assess progress and adjust plans.
- **Address Challenges Early:**
  - Identify bottlenecks or risks and take action to resolve them.
- **Collect Feedback:**
  - Seek input from your team and stakeholders to refine your approach.

Step 6: Close the Project

- **Wrap Up Deliverables:**
  - Ensure all outputs meet the defined standards.
- **Conduct a Retrospective:**
  - Review what worked, what didn't, and what can be improved.
- **Celebrate Successes:**
  - Acknowledge individual and team contributions to build morale.

Essential Tools for Every Step

- **Communication:** Slack, Zoom, Microsoft Teams.
- **Task Management:** Trello, Asana, ClickUp.
- **Planning:** Gantt charts, Kanban boards, Google Sheets.
- **Tracking:** Notion, Airtable, Monday.com.
- **Feedback:** Google Forms, Miro, post-project surveys.

Final Reminder

Start simple and scale as needed. Focus on the essentials, adapt your approach to fit your project, and remember: progress is built one task at a time. Use this checklist as your guide to move from idea to execution with clarity and confidence. Let's get started!

2. Downloadable Resources

Templates, Tools, and Further Reading to Elevate Your Projects

This section provides practical resources to help you implement the strategies and techniques discussed throughout this book. From ready-to-use templates to recommended tools and curated further reading, these resources are designed to save you time and enhance your project management journey.

Templates

Download Link: Project Templates Pack
https://tinyurl.com/5xe9bn8f

The following free project templates are available for downloading at the link above.

1. Project Charter Template

A document to outline your project's objectives, deliverables, stakeholders, and timeline.

2. Kanban Board Template

A visual tool to organize and track tasks across stages (To-Do, In Progress, Done).

3. Risk Assessment Matrix

Identify, evaluate, and prioritize risks to stay proactive in managing challenges.

4. Weekly Check-In Meeting Agenda

A structured template to guide team meetings, ensuring productive discussions and updates.

5. Retrospective Summary Template

A simple format for reviewing what worked, what didn't, and how to improve after each project.

Tools

Here are some of the best tools for different aspects of project management. Many offer free tiers, making them startup-friendly.

1. Task and Workflow Management

- **Trello:** A Kanban-style tool to organize tasks and collaborate with your team.
  Trello
  https://trello.com/

- **Asana:** Ideal for tracking tasks, milestones, and deadlines in small or large projects.
  Asana
  https://asana.com/

2. Communication and Collaboration

- **Slack:** Instant messaging for teams, with channels for specific projects or topics.
  Slack
  https://slack.com/
- **Miro:** A digital whiteboard for brainstorming and visual collaboration.
  Miro
  https://miro.com/

3. Planning and Documentation

- **Notion:** Combines note-taking, task management, and project tracking in one tool.
  Notion
  https://www.notion.so/
- **Google Workspace:** Includes Docs, Sheets, and Slides for collaborative planning and documentation.
  Google Workspace
  https://workspace.google.com/

4. Analytics and Reporting

- **Tableau:** A powerful tool for data visualization and project reporting.
  Tableau
  https://www.tableau.com/
- **Airtable:** Combines spreadsheets with database functionality for tracking projects.
  Airtable
  https://airtable.com/

Further Reading

Expand your knowledge with these insightful books and articles, curated to complement the lessons in this book.

Books:

1. "The Lean Startup" by Eric Ries

- Learn how to test and iterate quickly in startup environments.
  Amazon Link
  https://tinyurl.com/48sxnu6b

2. "Deep Work" by Cal Newport

- A guide to cultivating focus and productivity in a distracted world.
  Amazon Link
  https://tinyurl.com/yckd88wb

3. "Radical Candor" by Kim Scott

- Master the art of honest, empathetic communication and feedback.
  Amazon Link
  https://tinyurl.com/29j4bdhp

4. "Scrum: The Art of Doing Twice the Work in Half the Time" by Jeff Sutherland

- A practical introduction to Agile project management principles.
  Amazon Link
  https://tinyurl.com/6mnzjcxs

Articles:

1. "How to Conduct a Successful Retrospective" (Atlassian Team Playbook)

- Step-by-step guidance on running effective post-project reviews.
  Atlassian Playbook
  https://tinyurl.com/4nsxt8eb

How to Use These Resources

1. **Download and Customize Templates:** Start with the templates provided and adapt them to fit your project's unique needs.
2. **Test New Tools:** Experiment with one or two tools from the list to enhance your workflow.
3. **Dive Deeper with Reading:** Explore the recommended books and articles to refine your understanding and skills.

The Takeaway

These downloadable resources, tools, and further reading materials are your companions on the journey to becoming an exceptional project manager. Use them to start your next project with clarity and confidence, and don't hesitate to adapt them to your unique challenges. The future of your success is in your hands - let these resources guide the way!

3. Reflection Prompts

Questions to Internalize Key Lessons

Reflecting on what you've learned is a powerful way to internalize concepts and make them actionable. These chapter-specific prompts are designed to help you think deeply about the key lessons in this book and how to apply them to your projects. Take a few moments after each chapter to answer these questions, either in writing or discussion with your team.

Chapter 1: Understanding Project Management for Startups

1. How do the constraints in your specific startup (limited resources, time, budget) currently impact your approach to projects? Which constraints could you transform into strengths?
2. Think about a recent project that didn't go as planned. Which traditional project management assumptions or practices may have hindered your progress rather than helped?
3. How might simplifying your current project management approach lead to better outcomes? Which complex processes or tools could you streamline without losing effectiveness?

Chapter 2: Laying the Foundation for Success

1. What are the primary objectives of your current or next project? Are they SMART (Specific, Measurable, Achievable, Relevant, Time-bound)?
2. Consider your most recent project charter (or lack thereof). What essential elements were missing that, in hindsight, would have helped align stakeholders and clarify expectations?
3. How effectively do you currently involve stakeholders in the early stages of projects? What specific stakeholder could you better engage in your next initiative?

Chapter 3: Choosing the Right Approach for Your Project

1. Which aspects of the different methodologies (Agile, Waterfall, Kanban, etc.) have been most effective for your specific team's working style and projects? How might you customize these approaches further to better fit your specific needs?
2. When have you relied on intuition to guide a project decision? What was the outcome, and how could you better balance intuition with process in the future?

3. How flexible is your current methodology? What signs would indicate it's time to adapt your approach mid-project?

Chapter 4: Essential Tools and Templates

1. Which of your current project management tools adds the most value, and which might be creating unnecessary complexity without proportional benefits?
2. Think about your team's current workflow. Where do digital tools enhance your process, and where might analog methods (whiteboards, sticky notes, paper) actually be more effective?
3. Which templates do you currently use repeatedly? How might you refine them to better suit your specific needs and save even more time?

Chapter 5: The Startup Project Cycle: A Step-by-Step Guide

1. Which phase of the project cycle (initiation, planning, execution, monitoring, closure) does your team handle least effectively? What specific improvement could you make to improve your overall results?
2. How effectively do you monitor and adjust projects once they're underway? What early warning signs of problems do you often miss?
3. How thorough is your project closure process? What valuable insights might you be failing to capture by rushing through or skipping this phase?

Chapter 6: Managing Small Teams and Limited Resources

1. What's one task you're currently handling that could be more effectively delegated to develop a team member's skills while freeing your time for higher-value work? How do you currently determine which tasks to delegate versus handle yourself? What criteria could you use to make better delegation decisions?
2. What's one resource constraint (time, money, expertise) that consistently challenges your projects, and what creative approach could you try to work within or around this limitation?
3. If you manage remote or hybrid teams, what challenges emerge most frequently? How might you adapt your communication and management style to address these challenges and better support team members who may feel disconnected?

Chapter 8: Dealing with Uncertainty and Pivots

1. Reflecting on the Hidden Inflection Point Theory, can you identify projects where you've experienced this 60-70% momentum drop? How did you handle it, and what would you do differently now? What specific

leadership adjustments could you plan for this point in your current project?

2. What signals or metrics would indicate to you that a project needs a pivot rather than just minor adjustments? How could you make these signals more visible to the entire team?
3. How do you personally cope with the stress of uncertainty? Which strategies from this chapter align with your strengths and which would require more deliberate practice?

Chapter 10: Real-World Startup Project Stories

1. Which of the success stories most resonates with your current situation, and what specific lesson could you apply immediately?
2. Looking at the failure stories, which warning signs do you recognize in any of your current projects, and how might you address them before they become critical?
3. How effectively do you document and share your own project stories (both successes and failures) to build a learning culture within your organization? What system could you create to better institutionalize this knowledge?

Chapter 11: The Leadership Mindset for Project Success

1. How would your team describe your communication style as a project leader? What one aspect could you improve to create more clarity and alignment?
2. When team collaboration breaks down, what are the typical root causes in your organization? How might you address these systemic issues rather than just the symptoms?
3. Which resilience-building practice from this chapter seems most valuable for your current leadership challenges? How will you incorporate it into your routine?
4. What's one way you could foster greater psychological safety within your team to encourage more honest conversations about project challenges?

Chapter 12: Taking Your Skills to the Next Level

1. What does "continuous improvement" currently look like in your organization? Is it systematic and intentional, or more ad hoc and reactive?

2. Which emerging trend in project management (AI, remote collaboration, data-driven decision making) presents the greatest opportunity for your specific context? Which pose the greatest threat?
3. Of all the concepts presented in this book, which one will you implement immediately? What specific actions will you take in the next week to apply this knowledge?

Conclusion: Your First Project Management Success Story

1. What's the smallest, simplest project management improvement you could implement tomorrow that would demonstrate the value of a more structured approach?
2. Which tool or template will you apply immediately, and to which specific project or process?
3. What personal success metric would tell you that you've grown as a project leader six months from now? What specific improvements do you hope to achieve?

How to Use These Prompts

1. **Write It Down:** Take a few minutes after each chapter to jot down your answers. Revisit them periodically to track your growth.
2. **Discuss with Your Team:** Use these prompts in team retrospectives or meetings to spark meaningful conversations and align goals.
3. **Review and Reflect:** At the end of the book, review your responses to see how your understanding and approach to project management have evolved.

The Takeaway

Reflection is a powerful tool for turning knowledge into action. These prompts are designed to help you think critically about your growth as a project manager and leader, ensuring you don't just learn the lessons but live them.

Take the time to reflect - your future projects will thank you for it.

Made in the USA
Monee, IL
04 November 2026

67381288R00118